APPLE WATCH FOR SENIORS

The Most Complete Easy-to-Follow Guide to Master Your New Apple Watch. Unlock All Its Features with Step-by-Step Illustrated Instructions and Useful Tips and Tricks

Gary Watts

Table of Contents

Introduction

Apple is hailed as the powerhouse of consumer electronics design and user experience, but it is also an unexpected driver of modern fashion. When the iPod launched a decade ago, it was notable for storing and playing over 1,000 songs on a tiny device the size of a deck of cards. However, one of its lasting impacts was the introduction of white earbuds to almost every public space. Today, wearing them is so commonplace that we never even consider it. Yet, just ten years ago, they were a fashion statement screaming, "I own an Apple product."

With the recent release of the Apple Watch, the company strives to make another fashion statement: "The wearable personal technology we have seen in movies can now be even more personal, more distinctive, and more stylish."

The Apple Watch is not a sliver of glass you keep in your pocket or bag; it is essentially a timepiece with an integrated communications center and a connection to your personal information and devices that you wear all the time and anywhere.

You may currently be wearing a wristwatch, or if you are like many people, you haven't worn one for years since your last piece was replaced by the massive screen smartphone in your pocket. What is the appeal of the Apple Watch as a timepiece?

The Apple Watch is essentially an extension of the data you deal with daily, and a shortcut device to access it. Without making the distracting motion of getting your iPhone out of your pocket or dashing to the next room to get it, you can keep your schedule, notifications, and reminders conveniently with you at all times. You can also interact with people via phone or text messages, reach out to other Apple Watch owners via the Digital Touch features which helps keep track of various fitness goals and record vigorous exercise information generated by your iPhone's and watch' sensors and safely and conveniently pay for your shopping using Apple Pay and much much more.

This book is your simplified and condensed guide to the Apple Watch and its possibilities and capabilities. If you already have the watch, you will agree that owning it is initially a delightful and surprising experience and, later a perplexing learning process as you discover its capabilities and learn how to solve problems with its help.

The buzz that Apple Watch has created in the tech world right now is hard to ignore.Cupertino's very first wearable gadget looks to become the next big chapter in the history of computing by bringing all your favorite apps, tools and notifications to your wrist. With all these capabilities in mind, we are bringing you this book to help you master the ins and outs of your Apple Watch, and get to know what is inside and how you can use it to simplify your life.

Should you buy an Apple Watch?

Apple Watch is a revolutionary new product with long-established roots in technology and time-keeping. However, most people are confused as to whether spending over $350 on a wrist device they don't really understand is worth it. If you purchased this book because you want to know more about the watch before you buy or to help you decide whether to purchase one or not, you made the right move.

Everyone today needs a phone. Most people today need a computer. Much like the iPad, the Watch feels like an extra accessory, making it difficult to figure out whether you need it or not. To make the decision process a little bit easier, let's break it down.

Whether you should buy an Apple Watch or not boils down to how compelling the main features are for your lifestyle – by themselves or when combined. These features include timekeeping, health and fitness tracking, informational and notifications widgets, remote control, Apple Pay, and communication.

Simply put, the Apple Watch is the shuttlecraft to the iPhone's starship. Most of the activities the watch is capable of can already be done on your iPhone; the watch merely makes them more convenient. Only a select few are exclusive to the Apple Watch.

This essential Apple watch guide will save you the time and effort of investigating with trial and error. In next to no time, you'll be exploiting the full capability of your new gadget and making your daily life a lot easier with a few simple touches of a screen.

Let's get started!

Chapter 1:

What's New?

The Latest Apple Watch

The Apple Watch Series 7, unveiled in September 2021 and replaced the Apple Watch Series 6, is the latest iteration of the Apple Watch, which was first released in 2015 and superseded the Series 6. It is based on the design of previous Apple Watch models with a more rounded design and offers some noteworthy new features, including larger displays, longer life, and faster charging.

It is available in new 41 and 45mm size options, which are 1mm larger than previous generations' 40mm and 44mm options, and the cases, have been finished with softer, rounded edges. The Apple Watch Series 7 models include a black ceramic and sapphire crystal case back and a digital crown with tactile feedback, similar to the Apple Watch Series 6. The Digital Crown has an integrated sensor for ECG measurement.

The new models have a larger and redesigned retina display with more screen real estate thanks to narrower bezels. The display has a unique breaking edge that almost curves towards the case. There are UI enhancements, and two unique watch faces to take advantage of the larger displays. The low-power OLED Always-On Display (LTPO) technology debuted with the 5 Series is carried over to the 7 Series, allowing consumers to see the dial and complications at all times

With a crack-resistant front glass, IP6X dust resistance, and WR50 water resistance, the Apple Watch Series 7 is more durable than prior versions.

Apple Watch Series 7 models can also charge up to eight hours of sleep tracking time in only eight minutes of charging.

It offers the same health capabilities as previous versions, including blood oxygen monitoring, Electrocardiogram (ECG), sleep monitoring, fall tracking, and detection of loud sounds. It also supports Apple Pay transactions and emergency calls using SOS.

Apple offers the 7 Series with both GPS and GPS + LTE models. LTE Apple Watch models can be managed via LTE without an iPhone nearby.

Midnight, Starlight, Green, Blue, and (PRODUCT) RED are among the five new aluminum body colors available. Silver, graphite, and gold are still available in stainless steel, while silver and space black is still available in titanium. Apple continues to sell Apple Watch Nike versions in aluminum and Apple Watch Hermès models in stainless steel.

WatchOS 8

The Apple Watch runs an operating system called watchOS, and the Apple Watch Series 7 has watchOS 8 installed. The watchOS 8 update brings additional capabilities to enable clients to stay healthful, energetic, and connected with colleagues and family, with most of the latest components building on the improvements made in iOS 15.

The wallet has been updated to include ultra-broadband support for digital vehicle keys as well as additional digital keys for opening doors at home, work, and hotels. These key new features work with Apple Watch's tap-to-unlock feature. In some states, users can add their driver's license or status ID to Wallet, and select TSA checkpoints can start accepting digital IDs.

The Home app has been redesigned to make accessing HomeKit accessories and scenes easier when needed, with status updates for thermostats, light bulbs, and other accessories. HomeKit gadgets can now be managed from within the room, and users with HomeKit-enabled cameras can now see who's at the door from their wrists. There is a quick touch feature for intercom users to get in touch with everyone in the house.

Apple has added two new workout types, Tai Chi and Pilates, which can be selected when selecting a workout on the Apple Watch. For Apple Fitness + users, picture-in-picture support, filter options, and options to stop and continue an ongoing workout on any device.

The Breathe app is now the Mindfulness app and has been enhanced with a new Breathe experience and reflection session for conscious intent. Reflect offers users a thoughtful idea that encourages a positive attitude. The Breathe and Reflect experiences feature new animations and a variety of meditation tips.

During sleep, the Apple Watch now measures breathing rate (the number of breaths per minute), sleep time, heart rate, and blood oxygen. Breath data can be viewed in the Health app and is a metric that can be used to track overall well-being.

There's a new vertical clock face that extracts portrait photos from the iPhone and uses depth data to superimpose the time on favorite people's faces. The Photos app has been redesigned with new ways to view and navigate in collections. Trending reminders and photos are synced to the Apple Watch and can be shared right from your wrist.

Apple has added a special Find Items app to help you find your lost devices and redesigned the Music app so that users can share songs, albums, and playlists. Bad weather notifications, precipitation alerts for the next hour, and updated complexities are all available in the Apple Watch Weather app.

Scribble, dictation, and emoji may all be mixed in a single message in the Messages app, and there's a new option to modify spoken text. With watchOS 8, you can send GIFs in messages to your Apple Watch, and there's a new Contacts app that makes it simple to communicate with people when you don't have access to an iPhone.

The new Focus function in iOS 15 syncs with the Apple Watch, allowing you to eliminate distractions and focus on the task at hand. Apple also suggests focusing modes. Then, when you train, you will be asked to select the Focus for Fitness option.

WatchOS 8 supports multiple timers simultaneously, and multiple apps support always-on viewing, including Maps, Mindfulness, Now Playing, Phone, Podcast, Stopwatch, Timer, and Voice Notes. Third-party developers can also create always-on views for their apps.

Apple added an AssistiveTouch feature that uses sensors built into the Apple Watch to detect hand gestures for control purposes.

Comparison With The Older Models

The 2020 and later models have an always-on watch face. Apple announced that Series 6 was 20% brighter than Series 5. The Series 7 is 70% brighter when your wrist is down compared to the Series 6.

The Series 6 and later have a blood oxygen sensor. The Series 6 S6 processor is 20% faster than the S5 processor in Series 5. Apple hasn't announced which processor is in the Series 7 at this time, so it's probably still the S6 SiP processor. Series 6 and Series 7 both have a U1 chip, which we'll look at in a bit.

The Blood Oxygen sensor employs four green, red, and infrared LED clusters to shine a light on your wrist. Four photodiodes on the back crystal of the Apple Watch measure light reflected back from blood vessels. The app measures blood oxygen levels in 15 seconds. If enabled, the app will take background readings throughout the day, usually when you're not moving or 'sleeping. Researchers are exploring if SpO2 could be an early sign of respiratory conditions like influenza or COVID-19.

Model	Processor	Comments	Bluetooth
Series 3	S3	Dual-core	4.2
Series 4	S4	64-bit dual-core, up to 2x faster than Series 3	5.0
Series 5	S5	64-bit dual-core, up to 2x faster than Series 3	5.0
Series 6	S6	64-bit dual-core, 20% faster than the Series 5	5.0
SE	S5	64-bit dual-core, up to 2x faster than Series 3	5.0
Series 7		Not available	5.0

The S4, S5, and S6 processors are up to two times faster than the Series 3 dual-core S3. Series 4 and later models use Bluetooth 5, which has more speed, better range, and lower power consumption when compared to Bluetooth 4.2 in Series 3.

In 2020, Apple added integration to set up an Apple Watch for a family member who doesn't have an iPhone. To set up a "Family" member's watch, you must be the family organizer or a parent guardian for a minor. Series 4 and later watches work with the "Family" setup. You use your iPhone for tasks like updating contacts, adjusting exercise goals, and adding photos. Family setup may not be a good solution if your family member is not local.

Model	Storage Capacity	Family Setup (Cellular Models)
Series 3	8 GB or 16 GB	No
Series 4	16 GB	Yes
Series 5	32 GB	Yes
Series 6	32 GB	Yes
SE	32 GB	Yes
Series 7	Not available	Yes

Apple plans to continue offering the Series 3 but plans to stop selling the Series 6 model.

Apple announced its plans to continue selling the Apple Watch Series 3 and the new Series 7.

The Apple ECG app arrived with the Series 4 model and is available on Series 5, Series 6, and Series 7. The app provides heart rate monitoring, similar to a single-lead electrocardiogram (EKG). The ECG app works by measuring your heart rate on your wrist while you touch the opposite hand to the electrode in the Digital Crown, creating a circuit.

Model	Accelerometer	ECG	Blood Oxygen Sensor
Series 3	Up to 16 g-forces	No	No
Series 4	Up to 32 g-forces with fall detection	Yes	No
Series 5	Up to 32 g-forces with fall detection	Yes	No
Series 6	Up to 32 g-forces with fall detection	Yes	Yes
SE	Up to 32 g-forces with fall detection	No	No
Series 7	Up to 32 g-forces with fall detection	Yes	Yes

If the Apple Watch detects a significant, hard fall while you're wearing your watch, it taps you on the wrist, sounds an alarm, and displays an alert. Apple changed the algorithms in 2021 to identify when you fall during a workout like dance or cycling. Your watch will automatically contact emergency services if you do not respond to the prompt. Recently, Dr Sumbul Desai said falls are one of the most common reasons to go to the ER across all age groups. We all can benefit from the fall detection feature.

Apple focused on durability in 2021, and the Series 7 has Ip6X certification for dusty environments and the WR50 certification for water resistance to 50 meters. The redesigned front crystal is crack resistant and 50% thicker than the Series 6, and has a flat base.

Apple made a fundamental design change in the Series 7 Retina display by increasing the size of the chassis and reducing bezels. The re-engineered display reduces borders by 40% to provide 20% more screen area than the Series 6. The touch screen is integrated into the always-on OLED panel and provides 20% more screen area than Series 6.

The buttons in apps like Stop Watch, Activity, or Alarms are larger. In apps like Messages, you'll notice 50% more text on the screen, and you can tap, type or swipe to enter text. The Messages app also has a full A-Z QuickPath keyboard. The new watch face "Dynamic Contour" takes advantage of the improved display with numbers that appear to wrap around the edge of the screen.

Model	Screen	Always-on Display	Comments
Series 3	OLED	No	
Series 4	LTPO OLED	No	30% larger display areas than Series 3
Series 5	LTPO OLED	Yes	30% larger display areas than Series 3
Series 6	LTPO OLED	Yes	2.5x brighter than the Series 5, and 30% larger display areas than Series 3
SE	LTP OLED	No	30% larger display areas than Series 3
Series 7	LTPO OLED	Yes	20% larger display than Series 6

Titanium is 45 percent lighter than stainless steel with twice the strength. The space black titanium finish has a diamond-like coating. Stainless steel cases are heavier and have a shiny appearance, while the aluminum has a matte finish. Apple says the Series 7 is now "recyclable aluminum," and the case is 40% thinner.

Model	Aluminum	Stainless	Titanium	Ceramic
Series 3	Yes	Yes		
Series 4	Yes	Yes		
Series 5	Yes	Yes	Yes	Yes
Series 6	Yes	Yes	Yes	
SE	Yes	Yes		
Series 7	Yes	Yes	Yes	

With Series 4 and later, the microphone location is on the other side of the watch, away from the speaker. This microphone location reduces echo noise for better sound quality than Series 3. The audio volume is also 50% louder than the Series 3, to accommodate the walkie-talkie app introduced in watchOS 5.

Both Series 4, Series 5, and Series 6 have 40mm and 44mm cases. The Digital Crown in Series 4 and later models have haptic feedback with the sensation of incremental clicks. Haptic feedback is not available in the Series 3 Digital Crown.

Model	Case Size	Speaker & Microphone	Comments
Series 3	38mm 42mm	1st gen, 50% louder than Series 3	Microphone/Speaker same side
Series 4	40mm 44mm	2nd gen, 50% louder than Series 3	Microphone opposite speaker
Series 5	40mm 44mm	2nd gen, 50% louder than Series 3	Microphone opposite speaker
Series 6	40mm 44mm	2nd gen, 50% louder than Series 3	Microphone opposite speaker
SE	40mm 44mm	2nd gen, 50% louder than Series 3	Microphone opposite speaker
Series 7	41mm 45 mm	Not available	Not available

Silver and Space Gray colors are available in all models, and the Series 5 gold color is also found in this year's models. Series 5 introduced Nike and Hermès options, also available in 2020. The Hermès has exclusive watch faces, and an elegant new band connector.

Model	Gold	White	Nike/Hermès	Midnight	Starlight	Red	Blue	Green
Series 3								
Series 4								
Series 5	Yes	Yes	Nike					
Series 6	Yes		Nike/Hermès	Yes		Yes		Yes
SE	Yes		Nike					
Series 7				Yes	Yes	Yes	Yes	Yes

The built-in rechargeable lithium-ion battery uses magnetic charging and lasts up to 18 hours. In 45 minutes, the new USB C cable charges the Series 7 battery, which is 33% faster than the Series 6. In 8 min, the battery is charged enough for 8 hours which is enough for sleep tracking.

The power-saving Bluetooth 5 and Low-temperature Poly Oxide (LTPO) display are two reasons for the longer battery life, along with the unique power-saving features of the X5 and X6 processors. The "Family Setup" Series 6 configuration, where the watch is not connected to an iPhone throughout the day, drops from 18-hrs to 14-hrs.

Model	Battery Life	Charging time
Series 3	18 hr	Full charge 2 hours
Series 4	18 hr	Full charge 2 hours
Series 5	18 hr	Full charge 2.5 hours
Series 6	18 hr	20% faster compared to Series 5. Full charge in 90 minutes
Series 6 Family Setup	14 hr	20% faster compared to Series 5. Full charge in 90 minutes
SE	18 hr	Full charge in 2.5 hours
SE Family Setup	14 hr	Full charge in 2.5 hours
Series 7	18 hr	Full charge in 45 minutes

The heart rate monitor, improved accelerometer, barometric altimeter, and gyroscope are ideal for health and fitness apps. The accelerometer can differentiate between a walk and a run and enables features like "Running Auto Pause" to identify when you're taking an exercise break. Apple claims the always-on altimeter measures change as small as one foot. You can review elevation data during outdoor workouts.

Model	Optical Heart Rate Sensor	Gyroscope	Magnetometer	Barometric Altimeter
Series 3	1st gen	Yes	No	Yes
Series 4	2nd gen	Yes - improved compared to Series 3	No	Yes
Series 5	2nd gen	Yes - improved compared to Series 3	Yes	Yes
Series 6	2nd gen	Yes - improved compared to Series 3	Yes	Always on
SE	2nd gen	Yes - improved compared to Series 3	No	Always on
Series 7				

The Series 4 and 5 models have a black ceramic back with a sapphire crystal and electrical heart rate sensors. The Series 3 does not have an electrical heart rate sensor. Radio waves easily pass through the front and back for better cellular service.

Unboxing The Apple Watch

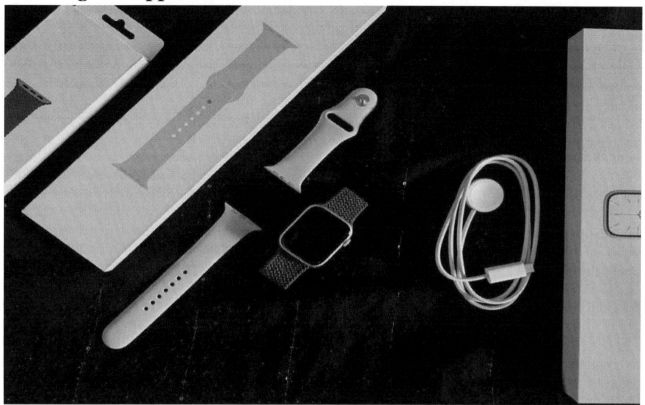

So you have just purchased your own Apple Watch, and now it's time to unbox it and get familiar with the design and interface. (If you have already started using your Apple Watch, feel free to skip to the next section). Upon opening the packaging, your Apple Watch is protected by a foam cover. You will also get a magnetic charging cable with a USB-C connection. The straps are provided separately.

You can remove your watch face from the foam cover and insert your straps by using the metal slides and inserting them into slots on the back of the watch face. You should hear and feel a click, letting you know that the straps have been inserted correctly. You can also choose which direction you want the clasps to face, and the watch can be worn on your right or left arm. It should be worn with the right fit—not too tight and not too loose. The back of the watch face needs to contact the skin for its features, like the heart rate sensor, to work properly.

The charging cable is very easy to use. Simply plug the USB-C into a power source and place the watch face's back onto the round surface. You will feel that it is held in place by magnets, and the watch should turn on or show an icon indicating that it is charging properly.

Never force your charging cables into their connectors, as the pins inside may bend or break. If you struggle to insert the cables properly, check for any obstructions or damage, and make sure you have matching components. Certain usage habits, like charging your device while in use, can cause the cables to become damaged, frayed, or break. If you repeatedly bend your cables in the same spot through repetitive use, they will become weak, brittle, and subject to breaking. Do not force the cables into sharp angles; instead, try to ensure that they can follow smooth curves. You should regularly inspect your cables and charging devices for any damage, breaks, or bends and discard them if you notice any problems.

Major Functional Parts Of The Apple Watch

It is very vital to familiarize yourself with the importance of all the external parts of the Apple Watch Series 7 to help you in operation and the general use of the watch.

The major functional parts of the watch you have to know are:

Digital Crown:

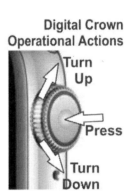

Digital Crown can be pressed to move from one interface to another and turned up and down to scroll through an open application.

- Digital Crown will enable you to return to the Home screen after you have navigated through different app's (application's) pages on your Apple Watch (It is also called Home Button).
- You can use it to scroll through all opened Apps.
- You can use it to access previously opened apps.
- You can use it to show Watch Face after editing.
- You can use it to access Siri.

Side Button

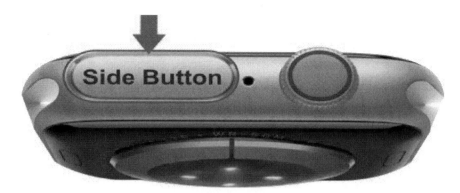

- Side Button can enable you to access your favorite contacts on your watch.
- It can enable you to send messages to your favorites.
- It can enable you to make use of the Power Reserve
- It can enable you to Power On or Power Off the watch.
- You can use it to access Apple Pay

Speaker and Air Vent

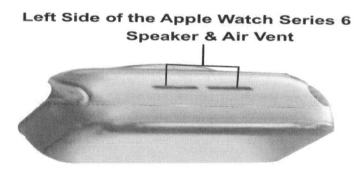

The left side of the Apple Watch Series 7 has two open spaces through which air enters and heat exits the Apple Watch, and the speaker to hear sound is positioned.

Microphone

The opening hole will enable your voice sound to reach the in-build Microphone that will convert your natural voice or external sound to electrical sound that can be recognized by the sound recognizing applications like Siri, Phone App, and Voice Recorder or Speech Dictation on your Apple Watch Series 7.

More so, the in-built speaker will enable you to hear the response of your speech request or for you to hear the ringing tone/alarm when a call or message or alert or notification enters.

Speaker can also enable you to hear audio sounds from your Apple Watch Series 7.

Ion-X Glass or Touch Sensitive Screen

- It will enable you to select your desirable app and move it around.
- It will enable you to edit or navigate any desirable feature on your watch with different types of Analog or Digital Clock patterns.
- The touch-sensitive screen will enable you to search, accept requests, or cancel unnecessary actions.
- You can swipe left, right, up, or down through your watch's touch-sensitive screen to locate essential features in the Applications or Control Center like "Do Not Disturb", Sleep Mode, Wi-Fi, Cellular … and many others. However, you will fully learn the general uses and practicals in this inevitable Apple Watch Series 7 Guide later.

Charging Face

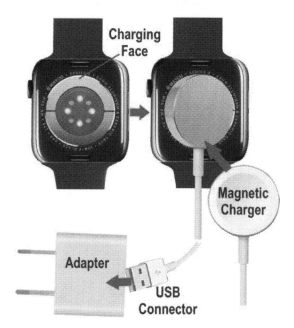

The charging face is at the back of the watch. All you need to do is to place the magnetic round charger at the back surface of the watch.

Connect the USB connector at the end of the charger cable with the Adapter and subsequently plug it into the electricity source.

It is very advisable that you carry out the charging connection in a fully ventilated location (i.e. where there is enough free-flowing of air).

Blood and Optical Heart Sensors

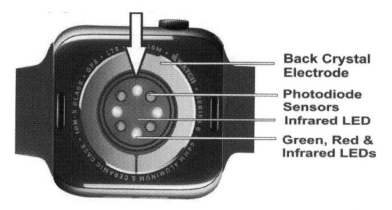

The two sensors enter into the wrist skin through black crystal showing green and red LEDs and infrared light (i.e. Photodiodes) to measure the quantity of oxygen and rate of the heartbeat. The quantity of light sent back into the watch is measured by the Photodiodes and transformed to data through the function of the sophisticated algorithms present in the watch.

The level of the red appearance tells the amount of oxygen in your body. If the red appearance is dark that means you have low Oxygen, but if the red displayed is light that means you have sufficient Oxygen in your body.

Chapter 2:

How To Set Up Your Apple Watch

Setting Up Your New Apple Watch

To set up and make use of your Smartwatch, you need an iPhone with the latest version of iOS installed inside. You should also ensure that your iPhone is Bluetooth enabled and connected to Wi-Fi or a mobile network.

1. Switch on your Watch and wear it on your wrist

To switch on your smartwatch, long-press the side button till the Apple icon shows on the screen. This may take a few minutes.

2. Bring your watch near your iPhone

Wait for the "**Use iPhone to set up this Apple Watch**" alert to show on your phone, and touch the **Continue** button. If this message does not appear on your iPhone's screen, just launch the Watch application on your iPhone, touch All Watches, and **Pair New Watch**.

If this is your watch, touch Setup for Myself. **Or, touch** Setup for a Family Member.

Ensure that your Watch and phone are close to each other until you finish the setup process.

3. Keep your Phone over the animation.

Position your iPhone so the watch face can be seen on the viewfinder. Wait for an alert that the Apple Watch is paired.

If you cannot utilize your iPhone's camera, or if you did not see the pairing message, touch **Pair Apple Watch Manually** and adhere to the guidelines on your display.

4. Set up as new or Restore from backup

If this is your first Apple Watch, click on **Set up as a New Apple Watch**. Or, touch the **Restore from Backup** button.

5. Read and Agree to the terms & conditions by tapping on the Agree button.

6. Log in with your Apple ID

If prompt, insert your Apple ID login code. If your Apple ID is not requested, log in later from the Apple Watch application: Touch General> Apple ID, then log in.

If Find My is not set up on your iPhone, you will be told to enable Activation Lock. If you see the Activation Lock screen, your Watch is already linked with your Apple ID. You must enter the Apple ID's email & password to continue setup.

7. Select your settings

The Apple Watch would show you the settings it shares with your Phone. If you enable features like Location Services, Find My, Diagnostics, & WiFi Calling for your Phone, these settings will automatically activate on your Apple Watch.

After that, you can decide to utilize other settings, like Siri & Route Tracking. If you have not already set up Siri on your Phone, it will activate after you select this option. You can also pick the text size for your smartwatch.

8. Generate a login code.

You can skip creating a login code, but you need one to use Apple Pay and some other features.

Touch **Add Long Passcode** or **Create a Passcode** on your Phone. Go to your Apple Watch to enter your new code. Touch Don't Add Passcode to skip this step.

9. Select Features and Applications

You are asked to add a card and set up Apple Pay. Then you will be walked through setting up features such as Activity, SOS, & automatic watchOS updates. You can set up cellular on cellular models.

You will now be able to install applications that the Apple Watch supports, or you can choose to install them later.

10. Allow your devices to synchronize

The syncing may take a long time depending on the amount of information you have.
Ensure your devices are near each other till you hear a chime & feel a tap from your watch,
then press the Digital Crown.

Monitor Daily Battery Usage

Monitor the power on your Apple Watch to ensure you are never stuck in a situation where it has run out of battery.

- To check how much power remains in the battery, open the control center by touching and holding the bottom of the screen and then swiping upwards.
- You will see an icon with a battery status indicator (%).
- Tapping this icon will open the battery menu, where you can turn on Power Reserve mode. If you have Bluetooth devices such as headphones connected, you can also see their remaining battery life here.

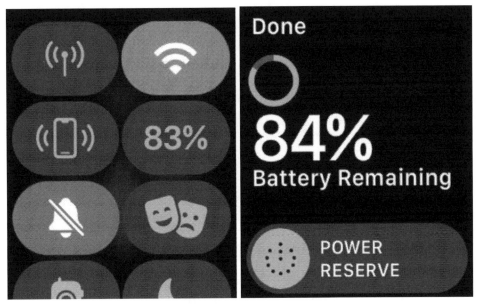

- You can also add a complication to your watch face that shows you your remaining battery life.

Prevent Background App Refresh

You can reduce power consumption on your Apple Watch by preventing apps from refreshing in the background. This prevents apps from updating or searching for new content while not in use.

- Open the Settings app and select General.
- Tap Background App Refresh, and turn this feature off.
- You can also turn this off for specific apps and not for others.
- Apps installed with complications on your watch face will continue to refresh even if Background App Refresh is turned off.

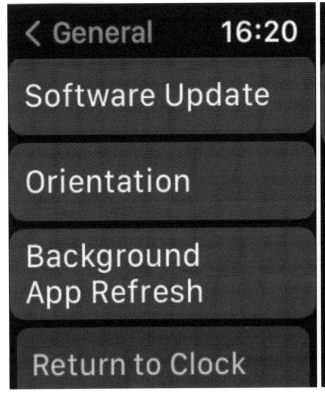

PowerReserve Mode

Your Apple Watch will automatically alert you when your battery level reaches 10% and remind you to enter Power Reserve mode. This feature is designed to help extend the life of your Apple Watch battery when it is low. It can give you a few extra hours of run time when you cannot charge your device, allowing you to see the time but preventing apps from running.

- To turn Power Reserve mode on, open the Control Center by touching and holding the bottom of the display and then swiping upwards.
- Touch the battery icon (%), and a new menu will appear where you can drag the slider to turn Power Reserve on.
- Your device will return to a normal power mode when you charge it.

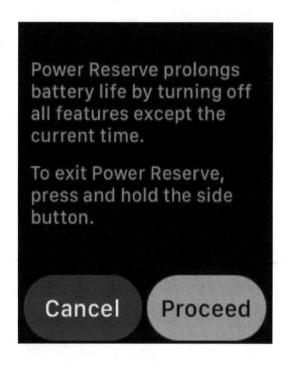

Customize The Watch Face

If you find a watch face you really like to use but prefer to see the time in a digital format rather than an analog one, you can easily customize it. You can also edit and customize many of the available watch faces. You can change the colors or select different data to display. You can also add or remove complications, the various data fields that display information such as weather, mindfulness, or activities.

1. Press the Digital Crown.

2. Touch and hold the display.

3. Swipe left or right to select the watch face you want to edit.

4. Tap the Edit button.

5. You will see different options depending on which watch face you are editing. Some watch faces allow you to edit many features, while others have limited options.

6. Swipe across the display to select different features to edit. Use the Digital Crown to make changes to each of the features.

7. To edit the complications, swipe all the way to the left. You will see each of the complications outlined by a border. Tap on the ones you want to edit, and use the Digital Crown to scroll through the available options.

8. To save your edits, simply press the Digital Crown.

You can easily create your own watch faces to suit your personal needs.

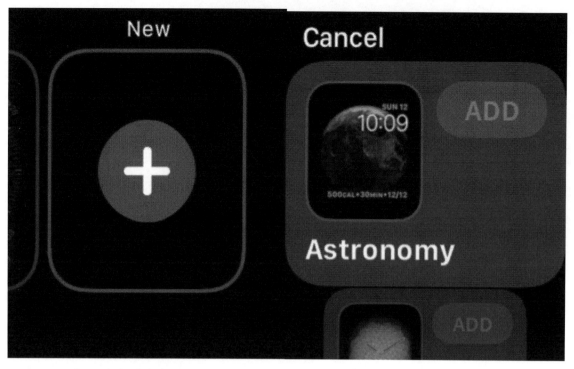

9. Press the Digital Crown.

10. Touch and hold the display.

11. Swipe all the way to the right until you see a [+] sign. Tap on this button.

12. Use the Digital Crown to scroll through all of the available templates.

13. Tap the Add button when you find one you like.

14. If customization options are available, you can scroll through them by swiping across the screen and using the Digital Crown to make the edits.

15. Press the Digital Crown to save your new watch face when you are done.

Configure Complications

You can add features -known as complications- to some watch faces so that you can quickly see things like the weather reports, stock prices, or other info from other installed applications.

- With your watch face showing, long-press your screen and touch the Edit button.
- Swipe to the left till you get to the end.

If a watch face has complications, you can find them on the last screen.

- Click on a complication to select it, then rotate your watch's Digital Crown to pick a new one - For example, Heart Rate or Activity.
- When you're done, press the Digital Crown to save your changes, and touch the face to move to it.

Turn to scroll through options.

Set Up And Use Notifications

One of the best features of the Apple Watch is the ability to see all of the notifications you would usually receive on your iPhone, including your messages, phone calls, emails, invitations, and other alerts from your apps. The notifications on your Apple Watch will be set to mirror the settings on your iPhone by default. This makes it much easier to stay in touch with your family and never miss any important communication.

Changing Notification Settings

Of course, not everybody wants notifications and communications on their wrist. Perhaps you want to receive notifications from your fitness and health apps but not your communication apps. You can easily turn this feature off or choose to see only selected notifications from certain apps or contacts.

- To change your notification settings, you will need to use your iPhone.
- Open the Watch app on your iPhone, then select Notifications.

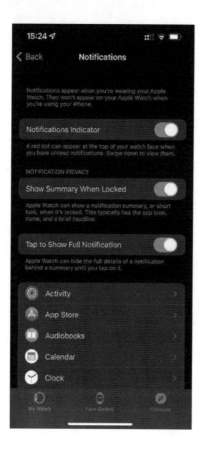

- You will see some options and then a list of all the compatible apps for your Apple Watch.

  1. Notifications Indicator: Toggle this setting to show or hide a red dot at the top of your Apple Watch's display when you have unread notifications.

  2. Show Summary When Locked: Toggle this setting to allow your Apple Watch to show a short summary of your notifications, including the app icon and name and a brief headline of the contents of the notifications when the device is locked.

  3. Tap to Show Full Notification: Toggle this setting if you would like to be able to tap a notification summary to open the full details.

  4. Show Notifications on Wrist Down: Choose to show notifications on your Apple Watch even when your wrist is down.

- Scroll through the list of apps and choose the notification settings for each of them.

  1. You will see the option to Mirror My iPhone or set Custom notification settings.

  2. The custom settings offered will differ for each app, but you will generally have the option to Allow Notifications, Send to Notifications Center, or turn Notifications Off. **Remember, these changes will only affect your Apple Watch, and your iPhone notifications will remain the same as before.**

3. You can also change the alert settings for the notifications by choosing to turn sounds and Haptics on or off.

Changing Notification Settings on Your Apple Watch

You can access a limited number of notification settings directly on your Apple Watch.

- When you receive a notification, swipe left on it and then tap it to bring up a small menu where you can choose to:

   1. Mute for 1 hour: This action will mute all notifications on your Apple Watch for the next hour. You can still see these notifications on your iPhone.

   2. Mute for Today: This action will mute all notifications on your Apple Watch for the rest of the day. You can still see these notifications on your iPhone.

   3. Add to Summary: All future notifications from this particular app will be sent straight to the summary on your iPhone.

   4. Turn off Time Sensitive: Prevent Time-Sensitive notifications from alerting you on your Apple Watch. These notifications are a special feature associated with Focus, and you will receive them even if you are in a Focus mode.

   5. Turn off: This action will stop all notifications from the particular app from appearing on your Apple Watch. To turn these notifications back on, visit the Watch app on your iPhone.

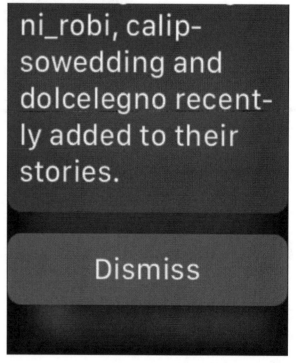

Make Use Of Limited Storage

Your Apple Watch comes with a total storage space of 32GB. To check the storage capacity left directly on your Apple Watch:

- Proceed to **Settings**.
- Click **General**.

- Then click **Usage**.

The **Storage** section displays the amount of available space and the amount of used space.

Scroll down the screen to see how much space has been used and the remaining space left.

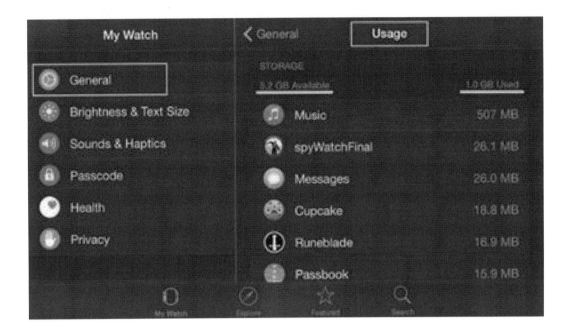

Checking Total Space from Your iPhone

Total storage space check can also be done on your iPhone to do this:

- Open **Apple Watch** app on your phone.
- In the **My Watch** section, proceed to **General**, then click **About**.

Wait for a few moments, and the total storage capacity and available storage space will be displayed. The number of media files you have on your watch, such as **Songs, Photos** as well as the number of applications on the device, will be displayed.

Removing Applications from Watch App

- Launch **Apple Watch** app on your iPhone.
- Tap on **My Watch** in the bottom navigation if you are not there already.
- Scroll down to the installed apps section.
- Find the app name you wish to remove from your Apple Watch, then tap on it.
- Toggle off the option for **Show App** on Apple Watch.

Deleting Music from Your Apple Watch

- Open **Apple Watch** app on your iPhone.
- Proceed to **My Watch** tab at the bottom of the screen and tap on it.
- Swipe to **Music** and tap it.

- Tap on **Edit** in the upper-right corner.
- If there is no **Edit** option, that means no music has been synced to your watch.

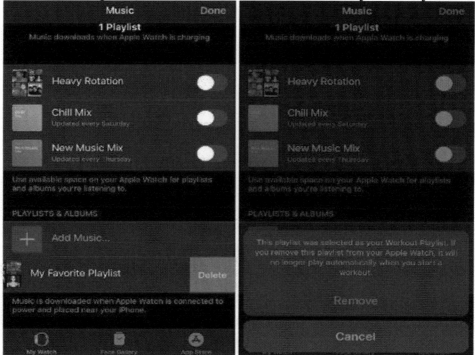

- Proceed to the **Playlists and Albums** section.
- Find the music you wish to delete and tap on the **Delete** next to it.

You can also turn off any automatically added playlists that you do not want on your Watch.

Control Your Music On Your Wrist

You may listen to Apple Music on your managed Apple Watch as long as you're connected to cellular/Wi-Fi if you're a member of a Family Sharing group with an Apple Music family membership.

- On your controlled Apple Watch, open the Music app and then do one of the following actions.

   1. Select Library to access music stored on your Apple Watch.

   2. Tap Listen Now to hear music that has been specially selected for you based on your listening habits.

   3. Tap Search, then enter, dictate, or doodle an artist, album, or playlist (Apple Watch Series 7 only).

   4. Please keep in mind that Scribble is not accessible in all languages.

   5. Select a playlist curated by Apple Music editors for children and teenagers.

   6. Select albums or playlists from your Apple Watch.

- Use the Music and Now Playing apps' music controls to play and choose music.

Use The Watch With An iPhone

- Wrap your wrist around your Apple Watch. Adjust the band or choose a band size that fits your wrist snugly yet comfortably.
- Just push and hold the side button to activate your Apple Watch until the Apple logo appears.
- Wait for the Apple Watch pairing screen to display on your iPhone with the Phone near the Watch, and then touch Continue.

  Or, locate and launch the Watch app on your iPhone and hit Pair New Watch.
- Select Create an Account for Myself.
- When asked, position your iPhone in such a way that your Apple Watch appears in the Apple Watch app's viewfinder. This establishes a connection between the two devices.
- Tap Set Up Apple Watch, then complete setup by following the on-screen directions on your iPhone and Apple Watch.

Tap Get to Know Your Watch to discover more about your Apple Watch while it is synchronizing. You can learn about new features on your iPhone, examine Apple Watch advice, and read this user guide. After configuring your Apple Watch, you may access this data by opening the Apple Watch app on your iPhone and clicking Discover.

Chapter 3:

Getting Started With Your Apple Watch

Accessing The Contacts List

The Contacts app ⊕ in your Apple Watch allows you to view, edit, and share your contacts from other devices that are signed in with the same Apple ID. It also allows you to create a contact and set up your contact card containing your information.

View Contacts On Your Apple Watch

- Go to the **Contacts app** ⊕ on your Apple Watch.
- Use (turn) the Digital Crown to browse through your contacts.
- Click on a contact to view its details.

Communicate With The Contacts App

Use the Contacts app to make calls, texts, emails, or start a Walkie-Talkie conversation.

- Go to the **Contacts app** ⊕ on your Apple Watch.
- Use (turn) the Digital Crown to browse through your contacts.
- Select a contact, then carry out any of the following:

    1. Tap ☎ to view the phone numbers of the contact. To make a phone call, tap a phone number.

    2. Tap 💬 to view an existing message thread or start a new one.

3. Tap ✉ to create and send emails.

4. Tap ◉ to start a Walkie-Talkie conversation or to invite a person.

Create A Contact

- Go to the **Contacts app** ◉ on your Apple Watch.
- Tap **New Contact** at the top of your screen.
- Enter the contact information (name or company, phone number, address, and email), then tap **Done**.

Share, Edit Or Delete A Contact

- Go to the **Contacts app** ◉ on your Apple Watch.
- Use (turn) the Digital Crown to browse through your contacts.
- Select a contact, scroll down, then click **Share Contact** > **Edit**, or **Delete Contact**.

Quickly Reach Your Favorite Contacts

To call a contact you've marked as a favorite in your iPhone's Phone app, tap Favorites, and then click a contact.

Calling A Friend From Your Watch

- On your Series 7 device, open the Phone app.
- Click **Contacts**, then scroll using the Digital Crown.
- To call a contact, tap the contact's name, then click the phone button.
- To initiate a FaceTime audio call, click **FaceTime Audio** or a phone number.
- Adjust the volume during a call by turning the Digital Crown.

To contact someone you've spoken with recently, click **Recents**, then tap a contact. To call a contact you've marked as a favorite in your iPhone's Phone app, tap Favorites, and then click on a contact.

On your series 7 device, you can also enter a phone number by following th steps listed below;

- Launch the Phone app when you access your series 7 device
- Enter the number on the Keypad, and then click the Call button.

Additionally, you can use the keypad to access extra digits while on a call. Simply press the More button twice, followed by the Keypad button.

Answering Or Rejecting Calls

Use the Phone app to make telephone calls and access other advanced features on your Apple Watch.

Answer A Call On Your Apple Watch

When you hear an incoming call, raise your wrist to see the caller.

- **Answer calls on Apple Watch**: Tap ◉ in the incoming call notification to answer the call and talk using a Bluetooth paired device or the Apple Watch's built-in microphone and speaker.

- Decline and send the call to voicemail**: Tap ◉ to decline the call.**
- **Answer the call on your iPhone instead**: Tap •••, then select an option. To switch to your iPhone, tap **Answer on iPhone**, the call will be placed on hold, and the caller will hear a repetitive sound until you answer the call on your iPhone. If you couldn't find your iPhone, touch and hold your Apple Watch screen's bottom edge, swipe up, then tap ◉.

Actions While On A Call

On your Apple Watch, if you are on a call that doesn't utilize FaceTime audio, you can do the following actions:

- **Quickly switch to your iPhone**: While speaking with the caller, wake your iPhone, then tap the green bar or button at the top of your screen.
- **Adjust the volume**: Use the Digital Crown to adjust the call volume. Mute your end of the call by tapping ◉.
- **Use the Keypad**: Tap ••• > **Keypad**, then enter the digits.
- **Switch to an audio device**: Tap •••, then select a device.

If you're on a FaceTime call, turn the Digital Crown to adjust the volume, tap ◉ to mute the call, or tap ••• and select an audio device.

TIP: You can quickly mute your Apple Watch when you get a notification by resting your palm on the watch display for at least three seconds. To activate this feature, go to the **Setting app** ◉ on your Apple Watch > **Sounds & Haptics**, then turn on **Cover to Mute**.

Listen To Your Voicemail

You receive a notification if a caller leaves a voicemail. To listen to the voicemail, click the **Play button** in the notification. If you want to check your voicemail later, go to the **Phone app** ◉ on your Apple Watch > **Voicemail**.

Sending Messages

- You can create and send texts on your series 7 using the Messages app, which supports text and images, Memoji stickers, emoji, and audio clips. Additionally, you can send money via Apple Pay and inform others of your location by including your spot in a message.
- ◉ **Ask Siri**. Say something like: "Tell Eve I'll be there in 40 minutes," then lower your watch to send it.

Create A Message

- On your Apple Watch, go to the **Messages app** ◉.
- Use the Digital Crown to scroll to the top of the screen > **New Message**.
- Tap **Add Contact**, select a contact in the list of recent conversations that appears, and then choose any of these options:

  1. Click ◈ to dictate a phone number or to search for a person in your contacts.

  2. Click ◉ to select from your full list of contacts.

  3. Click ◉ to enter a phone number.

- **Click** Create Message.

Reply To A Message

Use the Digital Crown to scroll to the bottom of the screen, then choose an option:

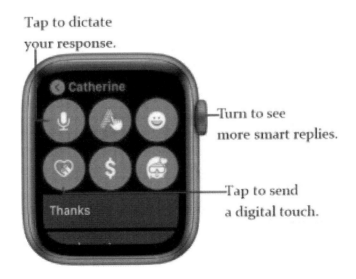

Tap to dictate your response.

Turn to see more smart replies.

Tap to send a digital touch.

You can also respond quickly with a Tapback (e.g., thumbs up or a heart) by double-tapping a specific message in a conversation.

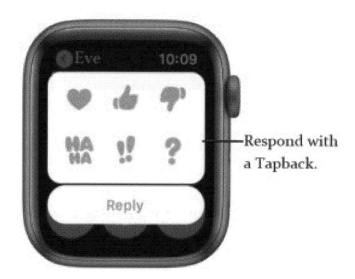

Respond with a Tapback.

Reply To A Specific Message In A Group Conversation

Replying to a specific message in a group conversation improves clarity and helps keep the group conversation organized.

- Touch and hold a message you want to respond to in a group conversation, then tap **Reply**.
- Write your response, then tap **Send**. The message will be seen by the person you replied.

Compose A Message

There are several ways you can compose a message on your Apple Watch:

- **Use the QWERTY and QuickPath keyboard**: Tap the letters or characters to input them, or use the QuickPath keyboard to slide from one character to another without lifting your finger. Lift your finger to end a word. Swipe up from the bottom of your screen to see the keyboard when it's not shown, then touch the Keyboard button.
- **Dictate the text**: Click 🎤, then verbalize the message. You can also verbalize the punctuations, e.g., "where do we meet the question mark." Tap **Done** when you are done.

Dictate the text

- **Use prepared response**: Browse through a list of helpful phrases you can use. Tap the one you want to use to send it. If you want to add your own phrase, go to the **Apple Watch app** on your iPhone > **My Watch** > **Mail** > **Default Replies**, then tap **Add Reply**.

- **Scribble the text**: Click ⊙, then use your finger to write the message. You can use predictive text options to turn the Digital Crown as you write. Tap Send when you are ready sending the message.

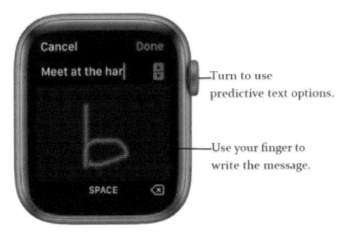

- **Send an audio clip**: Go to the **Apple Watch app** on your iPhone > **My Watch** > **Messages** > **Dictated Messages**, then click **Transcript, Audio,** or **Transcript or Audio**. Click **Audio** if you want the recipient to get your dictated text as an audio clip, not a text message. Click **Transcript or Audio** if you're going to choose the message format when you send it.
- **Add emoji**: Tap ⊙, select a category, then browse through available images. Tap the image you want to add to your message, and then tap **Send**.
- **Add a Memoji sticker**: Tap ⊙, select an image in your Memoji Stickers collection, then to send it, tap a variation.

- **Add a sticker**: Tap ⊙, turn the Digital Crown to scroll to the bottom, then tap More Stickers. Select one to send it. Use Messages on your iPhone if you want to see all your stickers or create new ones.

Send a text with your iPhone: If you've your paired iPhone close by while composing your text, you'll receive a notification on your iPhone, giving you an option to input the text with the iOS keyboard. Click the notification, then enter the text with the iPhone keyboard.

Sending Friends A Voice Message
- Open the phone app
- Make a call
- At the end of an unpicked call, select voice mail
- Make your voicemail
- Send

Reading And Processing Emails
You can read a mail on your Apple Watch, then write a reply using the QWERTY and QuickPath keyboard, scribbling, dictating, or using a prepared response. You can also quickly switch from your Apple Watch to your iPhone to write a response.

Read a new mail notification

- When you receive mail on your Apple Watch, just raise your wrist to read it.
- Swipe down from the top of the watch screen to dismiss the mail, or at the end of the mail, tap **Dismiss**.

If you didn't read the mail immediately, swipe down on the watch face to see your unread notifications, then tap the mail.

- To manage your Apple Watch email notifications, go to the **Apple Watch app** on your iPhone > **My Watch** > **Mail**, then tap **Custom**.

Read Your Mails In The Mail App
- On your Apple Watch, go to the **Mail app** ⊙.
    - Turn the Digital Crown to browse through the message list, then tap a message to read it.
Quickly Switch To Your iPhone
If you want to quickly switch to your iPhone to read the mail, do the following:
- Wake up your iPhone.
- Open the App Switcher on your iPhone (with Face ID) by swiping up from the bottom edge, then pause in the middle of the screen and lift your finger, or on iPhone models with a Home button, double-press the Home button to open the App Switcher.
Swipe right to browse the open apps, then tap the Mail button.

Replying To An Email

You can use your Apple Watch to create or reply to a message in the Mail app ⊙ and compose a message.

Create Or Reply To A Message

- On your Apple Watch, go to the Mail app ⊙.
- With the Digital Crown, scroll to the top of the screen, then tap **New Message**.
- To add a recipient, tap **Add Contact**, then tap **Add Subject** to create a subject line.
- Tap **Create Message** to compose your mail.

To reply to a message, use the Digital Crown to scroll to the bottom of a message you have received in the Mail app ⊙, click **Reply**, or click **Reply All** if there is more than one recipient.

Compose A Message

There are several ways you can compose a message on your Apple Watch:

- **Use the QWERTY and QuickPath keyboard**: Tap the letters or characters to input them, or use the QuickPath keyboard to slide from one character to another without lifting your finger. Lift your finger to end a word.
- **Dictate the text**: Click ↓, then verbalize the message. You can also verbalize the punctuations, e.g., "where do we meet the question mark." Tap **Done** when you are through.
- **Use prepared response**: Browse through a list of helpful phrases you can use. Tap the one you want to use to send it. If you want to add your own phrase, go to the **Apple Watch**

app on your iPhone > **My Watch** > **Mail** > **Default Replies**, then tap **Add Reply**.

- **Scribble the text**: Click ⊙, then use your finger to write the message. You can use predictive text options to turn the Digital Crown as you write. Tap Send when you are ready sending the message.

- **Send an audio clip**: Go to the **Apple Watch app** on your iPhone > **My Watch** > **Messages** > **Dictated Messages**, then click **Transcript**, **Audio**, or **Transcript or Audio**. Click **Audio** if you want the recipient to get your dictated text as an audio clip, not a text message. Click **Transcript or Audio** if you want to choose the message format when you send it.

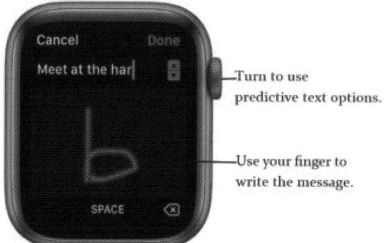

- **Add emoji**: Tap ⊙, select a category, then browse through available images. Tap the image you want to add to your message, and then tap **Send**.

Quickly Switch To Your iPhone

If you want to quickly switch to your iPhone to type the reply, do the following:

- Wake up your iPhone.
- Open the App Switcher on your iPhone (with Face ID) by swiping up from the bottom edge, then pause in the middle of the screen and lift your finger, or on iPhone models with a Home button, double-press the Home button to open the App Switcher.

Swipe right to browse the open apps, then tap the Mail button.

Mastering The Settings App
Always On Feature

The Always-On feature allows your device to show the watch face and time, even when you put your wrist down. When you raise your hand, your device will function fully.

- Enter the Setting application on your watch.
- Touch Display and Brightness, then click on Always on.
- Activate **Always On**, then click on the following options to configure them:
  1. Display Complications Data: Select the Complications that display info when you put your wrist down.
  2. Show Notification: Select notifications that appear when you put your wrist down.
  3. Show Apps: Select applications that remain visible when you put your wrist down.

Wake Your Watch Display

As a rule, you can wake up the Apple Watch display in the following ways:

- Raise your hand. Your watch would sleep again when you put your wrist down.
- Tap your screen, or click the Digital Crown.
- Rotate the digital crown upwards.

If you don't want the Apple Watch to wake up when you lift your wrist or rotate the Digital Crown, simply enter the Settings application on your Watch, head over to **Display and Brightness**, and then configure Wake on **Wrist Raise**.

Go Back To The Clock Face

You can decide how long before your Watch goes back to the clock face from an open application.

- Enter the watch's Settings application.
- Head over to General, touch Return to Clock, then scroll down and select the time you want your watch to go back to the clock face.

You can also press the Digital Crown to return to the clock face.

Wake Up To Your Last Activity

For some applications, you can configure the Apple Watch to take you back to where you were before the watch went to sleep. These applications include Workout, Mindfulness, Walkie-Talkie, Voice Memos, Stopwatch, Play Now, Podcasts, Timers, Music, Maps, & Audiobooks.

- Enter the Settings application on your Watch.
- Head over to General, touch **Return to Clock**, scroll down, and touch an application, then activate **Return to Application**.

You can also open the Watch application on your iPhone, touch the **My Watch** button, and then head over to General> Return to Clock.

Keep The Apple Watch Screen On Longer

You can keep your Watch screen on longer when you touch it to wake your watch.

- Launch the Settings application on your Watch.
- **Touch** Display and Brightness, **touch** Wake Duration, **and touch** Wake for 70 seconds.

Select A Language Or Region

- Launch the Watch application on your Phone.
- Touch **My Watch**, head to **General**, tap on **Language and Region**, touch the **Custom** button, and click on Watch Language.

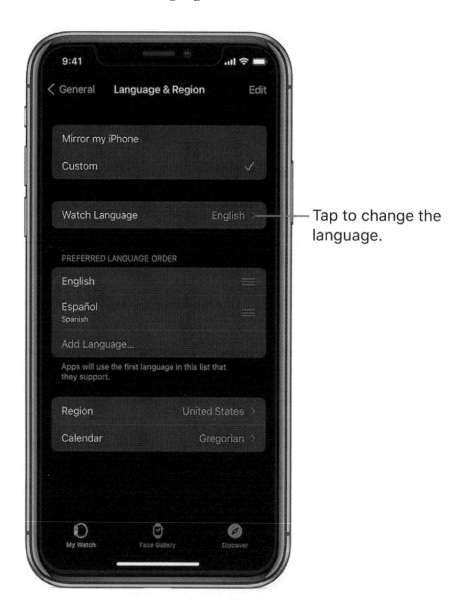

Tap to change the language.

Change Wrist Or Digital Crown Orientation

If you want to transfer your Watch to your other hand or if you prefer the Digital Crown on the other side, change your orientation settings so that raising your hand would wake your watch, and rotating the digital crown would move things in the direction you want.

- Launch the Settings application on your Watch.
- Head over to General, then touch Orientation.

You can as well launch the Watch application on your iPhone, touch My Watch, and head over to General> Watch Orientation.

Apple Watch control screen. You can customize your wrists and wristbands with Crown Digital.

Adjust The Text Size

You can adjust the text size that appears in any area compatible with Dynamic Views, such as the Setting application.

- Enter the Settings application on your watch.
- Head over to Display and Brightness> Text Size, then rotate the Digital Crown to change it.

Restart Your Watch

- Switch off your watch: Long-press your watch's side button until the sliders appear, then slide the Power Off slider to the right.
- Switch on your watch: Long-press the side button until the Apple symbol appears.

Force Restart Your Watch

If you cannot switch off your watch or if the problem persists, you can force restart your watch. This should only be done if you cannot restart your watch.

To force restart your watch, press & hold the Digital Crown & the Side button simultaneously for a minimum of 10 seconds till the Apple symbol shows on your screen.

Erase Your Watch & Settings

- Enter the Settings application on your watch.
- Head over to General> Reset, touch Erase All Content and Settings, and enter your code.

If you own a watch with a cellular plan, you have two options: Delete All and Keep Plan.

You can also launch the Watch application on your Phone, click on the My Watch button, head over to General> Reset, and click on Erase Apple Watch Setting & Contents.

If you cannot access the Setting application on your watch because you forgot the passcode of your watch, put your watch on its charger, and hold down the power button till Power Off is displayed on your screen. Long-press the Digital Crown and touch the Reset button.

Backup Your Watch

Your watch is automatically backed up to the phone it's paired with and can be restored from a backup. When you back your Phone to iCloud, Mac, or PC, the Apple Watch backup is added. If your backups are stored in iCloud, you won't be able to view their contents.

Update The Software Of Your Watch

You can update the Apple Watch software by checking for updates on your phone's watch application.

- Launch the Watch application on your Phone
- Click on My Watch, head over to General> Software Updates, and if there is an update, click Download & Install.

Security

- Open the Settings on your Apple Watch.
- Afterwards, click your [Your user name]
- Select Password & Security, and then perform one of the following actions:

  1. Change the password for your Apple ID: To change your password, tap Change Password and follow the on-screen instructions.

  2. Modify an app's or website's "Sign in with Apple" settings: Select an app by tapping Apps Using Your Apple ID. To disconnect your Apple ID from the app, tap Stop Using Apple ID. (You may be prompted to create an account)

3. Modify or add a reliable phone number: Click your current reliable phone number, verify it when prompted, and then click Remove Phone Number—if you only have one trusted phone number, you must first enter a new one. Click Add a Trusted Phone Number to add an extra trusted phone number.

4. Obtain a verification code for use on some other device or at iCloud.com: Select the Get Verification Code option.

Accessing the Dock

- To scroll through the Dock's apps, press the side button and then flip the Digital Crown.
- To open an application, tap it.

Turn the Digital Crown to see more apps. Tap one to open it.

App For The Dock

You may select to display the most recently used applications or up to ten of your favorite apps in the Dock.

- **To see recently used applications** on your iPhone, open the Apple Watch app, press My Watch, hit Dock, and then tap Recents. The most recently used application is shown at the top of the Dock, followed by subsequent applications in the order in which they were last used.
  Additionally, you may launch the Apple Watch's Settings app, ◎ hit Dock, and then touch Recents.
- **View favorite applications:** On your iPhone, open the Apple Watch app, hit My Watch, and then tap Dock. Select Favorites, then Edit, followed by a touch on ◉next to the applications you want to add. Drag ☰ to reorder them.

- **To remove an app from the Dock,** press the side button and then rotate the Digital Crown to the desired app. On the app, swipe left and then press/tap *X*.

- **Toggle between the Dock and the Home Screen:** Scroll to the Dock's bottom and hit All Apps.

Tip: Additionally, you may add complexities to your watch face with the applications you use the most. Consult Customize the watch face for further information.

Managing Your Watch With Control Center

The Control Center has a series of icons. Swipe up on the Apple Watch face to open the Control Center. Tap to toggle the options on or off.

- Cellular
- Wi-Fi
- Airplane Mode
- Battery
- Announce with Siri
- Find my (iPhone)
- Flashlight
- Focus
- Mute
- Sleep
- Theater Mode
- Water Lock
- Audio Output

Tip: Open Control Center from any screen. Touch the bottom of the screen until a semi-transparent preview of the Control Center appears, then swipe up.

Rearrange Icons In The Apple Watch Control Center

Follow these steps to customize your Control Center apps or set the app order.

- Swipe up on the Apple Watch face to open the Control Center.
- Swipe up and scroll to the end. Tap "Edit" to change items in the Control Center.

Focus

New in watchOS 8 and iOS 15 "Focus" expands "Do Not Disturb" and is designed to help you "Find Focus in the Day." You configure your "Focus" preferences in the "Settings" app on your iPhone or Apple Watch. When a custom "Focus" setting is active, the "Do Not Disturb" icon changes.

    Do Not Disturb
- Sleep (green bed icon)
- Personal (purple person icon)
- Work (blue contact card icon)

Filter notifications, signal to friends when you're not available, and hide distractions without missing what's important. As part of the "Personal" mode, you can choose the people and apps allowed to send you notifications when "Personal" is active. While setting up "Personal," the last setting is for "Time Sensitive" notifications, which includes alerts that your order is ready or your doorbell camera detected motion.

When one of the "Focus" modes is active on your iPhone, your Apple Watch will display an icon at the top of your watch face. So, for example, the "Personal" Focus mode is shown as a purple person icon.

Apple Watch Control Center Icons

Announce With Siri

watchOS 7 added, "Announce With Siri." ◻ This control is only available when supported AirPods or Beats are connected and a control button is added to the Control Center.

Cellular

When connected, the cellular status icon is green. When there is no connection, the status icon is grey.

Wi-Fi

When connected, the Wi-Fi status icon is blue. When there is no connection, the status icon is grey.

Airplane Mode

An orange airplane means Airplane Mode is active.

Battery

The battery status icon displays your battery level as a percentage. A red icon indicates your battery is low.

Tip: The status screen also displays the battery level of paired AirPods.

Do Not Disturb

🌙 Calls and alerts won't ring or light up the screen when "Do Not Disturb" is on. Alarms will still sound. The Do Not Disturb status icon on your watch face is a blue moon. You can continue a "Walkie-talkie" conversation if you turn on "Do Not Disturb," but other calls are silenced.

With watchOS 8, when you tap the Do Not Disturb icon, your "Focus" settings are displayed. Focus allows you to configure Do Not Disturb settings for Sleep, Personal, or Work activities in the Settings app on your iPhone, as shown in the next Chapter.

Find My (iPhone)

Find My (iPhone) may be the handiest feature if you tend to misplace your iPhone as frequently as do I! Swipe up on your watch face and tap the icon to instantly sound an alert on your companion iPhone. The blue icon has an iPhone with signal bars. At night, touch and hold the icon to flash a light on your iPhone.

Flashlight

The flashlight setting has three modes: the basic light, a strobe light, or a red light. Swipe to the left or right to choose your setting. When running at night, the strobe light is a nice safety feature. Press the Digital Crown to turn off the flashlight, or tap the icon in Control Center.

Silent Mode

Silent Mode will mute your watch. If you turn on Silent Mode while using the "Walkie-talkie" app, you can still hear chimes and your friend's voice.

Sleep Mode

Sleep Mode dims your screen and turns on Do Not Disturb.

Theater Mode

The picture of two masks is orange when Theater Mode is active. The screen stays dark and silent mode is also active until you tap the screen or press a button. When Theater Mode is active, your Walkie-talkie status is "unavailable."

Walkie-talkie

The walkie-talkie icon is a stylized walkie-talkie radio. The status icon appears after you create a connection with a contact. The icon is yellow when the walkie-talkie is turned on and indicates your status in the Walkie-talkie app is "available."

Water Lock

As I type this, I'm looking over my shoulder, expecting someone to say, "No way; you can't do that!" But this is straight from the horse's mouth (Apple being the horse) - you can go for a swim with your Apple Watch. Not only that, but the Workout app also has an option for "Open Water Swim" or "Pool Swim." "Water Lock" is automatically turned on when you start one of these workouts and locks the screen to avoid accidental taps.

The Activity app tracks "Swimming Distance" and "Swimming Strokes." Apps like "MySwimPro" are also fantastic for tracking swimming workouts.

Turn on Water Lock

- On the Apple Watch, swipe up from the bottom of the screen to open Control Center.
- Tap the water lock icon. It looks like a drop of water.

Turn off Water Lock

When your workout ends, turn the Digital Crown to unlock the screen and clear water from the speaker. Turn the Digital Crown until you fill the "blue circle" on the screen. When complete, an alert sounds, and the screen displays the message "Unlocked."

Audio Output

Use the Control Center's Audio Output to stream music or videos to your favorite speakers, AirPods, or headsets.

- Swipe up on the Apple Watch face to open Control Center.
- Tap the Audio Output icon.
- Tap "Connect a Device" and select the output device.

Tapping the audio output icon will also switch the audio output between paired Bluetooth devices.

Control Audio Volume

When playing audio, tap the audio status icon on your watch face, and turn the Digital Crown to adjust the volume. Control music, podcasts, or hearing aid volume.

Setting Up Do Not Disturb

All calls and notifications, except for alarms, will be silenced when Do Not Disturb (DND) is turned on. You'll notice a purple moon or DND icon come up at the top of the display.

- On your smartwatch, launch the Control Center.

- Touch the Do Not Disturb icon , and select an option.

Touch to manually toggle off Do Not Disturb or schedule it for an interval so that it automatically goes off. Make use of the Wallet app

Improving Sounds And Vibrations
How To Adjust The Sound

- Look for the Settings app on the Apple Watch.
- Touch Sounds & Haptic.

- Touch the volume control buttons under Alarm Volume or touch the slider, and then rotate the Digital Crown to adjust.

Or go to the Apple Watch app on your iPhone, tap Sounds & Haptics, and then drag the alert volume slider.

It can also reduce loud noises from headphones connected to your Apple Watch. In the Settings app, go to Sounds & Haptics> Headphone Safety, and then turn on Reduce loud sounds.

How To Adjust Haptic Intensity

You can regulate the power of the haptic or touch on the wrist used by e Apple Watch for notifications and alerts.

- Look for the Settings app on the Apple Watch.
- Touch Sounds & haptic and then turn on haptic alerts.
- Select Default or Explicit.

Or go to the Apple Watch app on your iPhone, tap My Watch, tap Sounds and Haptics, then select Default or Explicit.

Setting A Timer On Your Watch

You can monitor the time on your smartwatch by using the Timers app. And not just that, you can even set up multiple timers.

To set a timer using Siri, just say: "Set a timer for two hours."

Quickly Set a Timer

- On your smartwatch, launch the Timers app.
- To swiftly start a timer, touch the interval (such as 5, 3, or 1 minute). Otherwise, touch a timer you have previously used beneath "**Recents**."

- If you wish to create a custom timer, swipe down, and touch the **Custom** option.

- You can start a new timer with the same duration by tapping on **Repeat Timer** after the previous one has finished.

Pause/End a Timer

You can pause or end a timer once it's still running.

- On your smartwatch, launch the Timers app.
- Then, touch the Pause icon to pause.

- To resume the timer, touch the Play icon   .

- To end the timer, touch the End icon   .
- Create a Custom Timer
- On your smartwatch, launch the Timers app.
- Then, move to the top of the interface, and touch **Custom**.
- Next, touch seconds, minutes, or hours; proceed by turning the Digital Crown to change.
- Then, touch **Start**.

Tap hours, minutes, or seconds, then turn the Digital Crown.

Create Multiple Timers

- On your smartwatch, launch the Timers app.
- Proceed by creating and starting a timer using the steps outlined above.
- You can even use Siri to assign a tag like "Starbucks." Lift your smartwatch and say: "Set a 23 minutes Starbucks timer."
- Touch the "<" icon to go back to the Timers interface. Next, create and start a new timer.
- You'll notice the running timers will display on the Timers interface.
- To pause, touch the Pause icon.
- To resume, touch the Play icon.
- You can get rid of a running or paused timer that displays on the Timers interface, just swipe left and then touch "X."

Chapter 4:

Advanced Settings

Manage Daily Calendar Events

The Apple Calendar [29] app is the simplest way to display your schedule on your Apple device. Tap the Calendar complication (date) on your watch face to open the Calendar app at any time, or open the Calendar app with the side button. If you enable Family Sharing for iCloud, it creates a shared family calendar. watchOS 8 has an option to add Google accounts.

When you add a "location" to a calendar event on your Apple Watch, tap the address, and the "Maps" app opens with directions.

Add a Calendar Event

Use the calendar app on your iPhone to add an event to your calendar, as shown in Chapter 9, "Apple iPhone Apps, Calendar." You can also ask Siri to create a calendar event.

If you've enabled "handoff," you can quickly switch from your Apple Watch calendar app to the iPhone calendar app. Unlock your iPhone and tap the banner along the bottom of the screen to open the calendar app on your iPhone. Enable handoff in the Apple Watch app in the "General" settings screen.

Monitor Your Schedule

You'll configure all your sleep options in the "Full Schedule and& Options" screens. The additional "Additional Details" include the areas listed below. There are many options, including settings to show the time on your watch face during sleep mode, send "Sleep Reminders," and display "Sleep Results."

- Sleep Goal
- Wind Down (time)
- Wind Down Shortcuts
- Options

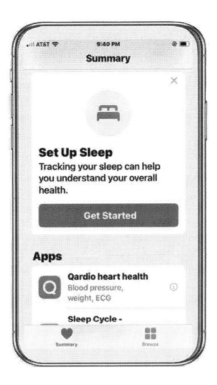

The "Full Schedule and Options" has the same alarm choices as "Your Schedule." The schedules you create here are based on the days of the week. In my case, I have "Weekdays" and "Weekends."

Basic Schedule Settings

The basic settings in "Your Schedule" or "Full Schedule" include the basic alarm options shown below.

- Wake Up Alarm (Enabled or Disabled)
- Sounds and Haptics
- Alarm Volume
- Snooze

In addition to the volume slider, "Sounds & Haptics" has vibrations and alarm sounds choices. You'll find the new "Bird song" alarm in this section.

To change the alarm time, drag the dial on the clock. My "First Schedule" is shown in the next figure. Tap to select the days of the week for this schedule. Blue is enabled, and white is disabled.

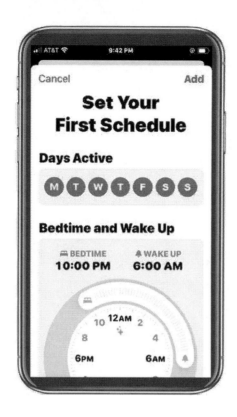

Add a Second Schedule

After creating your initial schedule, to add a second schedule for different days, tap "Add Schedule for Other Days." Tap the days that apply to this schedule. The active days are blue. Inactive days are white.

- Open the "Health" app on your iPhone.
- In the bottom left corner of the screen, tap "Summary."
- Swipe up and tap the arrow on the right side of the screen next to "Full Schedule & Options."
- Tap "Edit" underneath a schedule name.
- In "Alarm Options," swipe up to set "Wake Alarm." Toggle the switch on or off.
- In "Sounds & Haptics" at the top, set the "Vibration." The choices include "heartbeat" or "symphony."
- Choose an alarm sound. I think the default "Early Riser" is lovely, but I'd encourage you also to try "Helios," "Bird Song," or "Droplets."
- Toggle "Snooze" off or off.

Edit "Your Schedule"

"Your Schedule" refers to the next time the bedtime alarm is set. You might think of this as a temporary alarm for tonight.

- Open the "Health" app on your iPhone.
- In the bottom left corner of the screen, tap "Summary."

- Swipe up to the "Your Schedule" section, then tap "Edit."
- In "Alarm Options" swipe up to set "Wake Alarm." Toggle the switch on or off.
- In "Sounds & Haptics" at the top, set the "Vibration." The choices include "heartbeat" or "symphony."
- Choose an alarm sound. I think the default "Early Riser" is lovely, but I'd encourage you also to try "Helios," "Bird Song," or "Droplets."
- Toggle "Snooze" or off.

Get Directions On Your Wrist

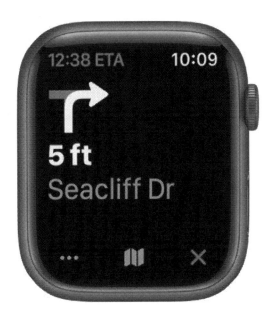

- Enter the Maps application on your watch.
- Rotate the Digital Crown to navigate to Recents, Guides, Favorites.
- Touch an entry to receive directions for cycling, transit, walking, or driving.
- Click on a mode to see the directions provided, then click on a route to start your journey and see an overview with turns, turn distances and street names.

Look in the upper left corner to see when you will arrive.

Explore The Music App

The Apple Music app can listen to songs, albums, playlists, or artists. The Music app was redesigned in 2020 and in 2021 includes the ability to share songs, albums, and playlists in Messages or Mail. There are two options for playing music: stream music over a cellular network or download music to your watch.

When you listen to audio on your iPhone or with CarPlay, the Now Playing app automatically opens on your Apple Watch. This feature can be disabled in the "Settings" app under the "General" category on your Apple Watch. Tap "Wake Screen" and then tap the toggle to disable "Auto-Launch Audio Apps."

The Play/Pause button includes a progress circle, and the AirPlay button quickly connects to headphones or speakers. The controls for the Music app are similar to the Podcast controls.

Play Music

The "Music" app and the "Now Playing" app allow you to control music from your Apple Watch. In the "Now Playing" app, you can control music on your Apple Watch and music on your iPhone or CarPlay.

- On your **Apple Watch**, press the Digital Crown.
- Swipe and tap "Music."
- Swipe up or turn the Digital Crown.
- Tap "On iPhone," "Now Playing," or "Library."
- Swipe and tap to select Playlists, Artists, Albums, or Songs.

- Tap the song to play.

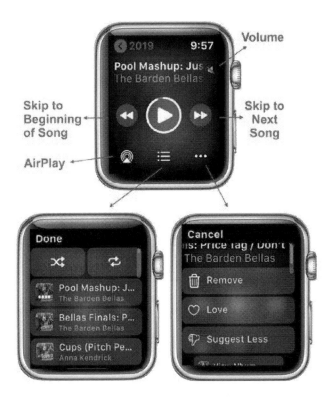

Add a Workout Playlist

You can automatically play music from a workout playlist whenever you start a workout. After watchOS 6, you can shuffle your workout playlist. Configure the playlist in the Apple Watch app on your iPhone.

- On your **iPhone**, open the Apple Watch app.
- Swipe to scroll down and tap "Workout."
- Swipe up and tap the "Workout playlist" option to select a playlist for your workouts.

Use Audio Output With Apple Watch

To stream music or videos to your favorite speakers, AirPods, or headsets, use the Control Center.

- Swipe up on the **Apple Watch** face to open Control Center.
- Tap the "Audio Output" icon.

In the "Now Playing" app, change the output destination by tapping the "Audio Output" icon in the lower-left corner of the screen. Tapping the audio output icon will also switch the audio output between paired Bluetooth devices.

Shuffle, Repeat, Source, and Output

When playing music at the bottom of the screen, tap the middle icon to see these options.

- Shuffle

- Repeat
- Source
- Playing Next

Auto Launch Audio Apps

Finally, when you swipe up, you'll see a list of watch apps. For each app, you can set a custom timer setting. So, for example, if you're using your Apple Watch to change slides in Keynote, you may want to set a longer timer before returning to the clock face.

- On your **Apple Watch**, press the Side Button.
- Tap "Settings," and then tap "General."
- Scroll down and tap "Wake."
- Tap "Auto-Launch Audio Apps."

Add or Remove Music on Your Apple Watch

To listen to music on the go when you don't have your iPhone with you, download albums or playlists to your Apple Watch. If you subscribe to Apple Music, the "Favorites Mix" and "New Music Mix" are automatically added.

- On your **iPhone**, open the Apple Watch app.
- Tap "My Watch," located in the left corner of the tab bar at the bottom of the screen.
- Swipe to scroll down and tap "Music."
- Tap "Add Music" and then tap the playlist or album you want to add. This is also where you could delete a playlist.
- To download music, connect your watch to Power, and place it near your iPhone.

Check Available Space

Music files can use up a lot of storage space. If you're wondering how much space music files are using, open the Apple Watch app on your iPhone to see detailed information.

- The count of songs on your watch.
- The count of photos on your watch.
- The number of applications on your watch.
- The total capacity.
- The available capacity.

  1. On your **iPhone**, open the Apple Watch app.

  2. Tap "My Watch," located in the left corner of the tab bar at the bottom of the screen.

  3. Swipe up to scroll down and then tap "General."

  4. Tap "About" to see available capacity. Swipe up to view Songs, Photos, and more.

Control Your iPhone Camera

The Apple Watch has a great camera 📷 remote control. You're probably thinking, "the Apple Watch doesn't have a camera," and that's true. Apple renamed the "Camera" app in watchOS 7 to "Camera Remote," which is more descriptive of what the app actually does. In addition to Apple's "Camera Remote" app, third-party apps like ProCamera, Hydra, and Camera Plus enhance camera remote control features.

If you don't see the Camera Remote app on your Apple Watch, check Content & Privacy Restrictions, as shown below.

Screen Time: Content and Privacy Restrictions

- On the **iPhone**, open the Settings app.
- Swipe up and tap "Screen Time."
- Swipe up and tap "Content and Privacy Restrictions."
- Tap "Allowed Apps" and enable "Camera."

Camera Remote and Timer

Position your iPhone to take a photo, then use the "Camera Remote" app on your Apple Watch to view a preview and take a photo. Note that your Apple Watch must be within 33 feet or 10 meters of your iPhone.

- On your **Apple Watch**, open the "Camera Remote" app.
- Position your iPhone to frame the shot, using your Apple Watch as a viewfinder.
- Tap the "Shutter" button.

View Your Photo Library

When browsing albums, tap a photo to select it. The "Photos Tab" has a new curated view for each day, month, and year. In the top right corner of the screen, there are buttons to mark the photo as a Favorite or Send, Delete or Edit the photo.

The Share 📤 button opens the Activity View, as outlined in Chapter 4. There are three tab bar rows in Activity View. The first row is "AirDrop."

The second row under AirDrop displays Share extensions. The Share extensions displayed relate to tasks specific to the current content. In the Photos app, share extensions include options like these.

- Share with Messages
- Share with Mail
- Shared Albums

The third row in the Activity View displays Action extensions, which include these options and more.

- Copy
- Slideshow
- Airplay
- Add to Album
- Use as Wallpaper
- Save to Files
- Assign to Contact
- Print

Tip: When the Activity View is open, swipe left and right in the tab rows to view more buttons.

Share Photos

The Photos ● app has been completely redesigned in watchOS 8 and automatically populates Memories, Featured Photos, and Favorites from your phone, as well as a Mosaic grid.

- Memory Highlights
- Featured Photos
- Favorites

When you open a "Memory," photos are displayed in a cool new mosaic layout. With watchOS 8, you can easily share photos through messages or mail.

- On your Apple Watch, open the Photos app and tap to select a photo.
- In the bottom right corner, tap the "Share" icon. Swipe up and tap either the Messages or Mail icon.

- When you swipe up, you'll see another option, "Create Face." Tap "Create Face" to create a new watch face from the photo.

To browse photos using the Apple Watch app "Photos," turn the Digital Crown. Follow the steps below to add or remove a photo library on your Apple Watch.

- Open the **Apple Watch** app on your iPhone.
- Tap "My Watch," located in the left corner of the tab bar at the bottom of the screen.
- Swipe to scroll down and tap "Photos."
- In the section "Photo Syncing" select photo albums.

The setting "Photos Limit" controls the number of photos on your watch. Only recent photos sync when the album size exceeds the limit.

Use Your Watch With Apple TV

Your watch can be used as a remote control for any Apple TV when connected to the same Wi-Fi network.

Connect Your Watch To Apple TV

If your iPhone is not connected to the WiFi network that your Apple TV was on before, join it now, then adhere to the directives below:

- Launch the Remote application ◎ on your watch.
- Touch your Apple TV. If you do not see it, touch Add Device.
- On your Apple TV, head over to the Settings application> Remote & Devices> Remote Application & Device, then choose Apple Watch.
- Enter the code shown on your watch.

When the pairing icon appears beside Apple Watch, the watch is ready to control your TV.

Control Your Apple Tv With Your Watch

Ensure your TV is awake, and adhere to the directives below:

Control another device.

10:09

SELECT

Swipe to move through Apple TV menu options; tap to select.

Tap to go back or touch and hold to return to main menu.

- Launch the Remote application on your watch.
- Select your Apple TV, then swipe right, left, down, or up to navigate through the Apple TV menu options.
- Touch to select the selected item.
- Touch the Pause/Play button to resume or pause what is playing.
- Touch the Menu button to go back, or long-press it to go to the main menu.

Disconnect & Remove The Apple TV

- On your TV, head over to the Settings application> Remote & Device> Remote Application & Device.
- Click on your Watch under Remote Application, and click on Unpair Device.
- Launch the Remote application on your watch and click the Remove button when the "lost connection" message appears.

Get More Active With Your Apple Watch

The Activity app on your smartwatch helps you monitor how much exercise, standing, and movement you're doing daily.

Add/Remove A Friend

If this is your first attempt at sharing an activity, you'll have to launch the Fitness app on your iOS device and click on "**Sharing**." Next, click "Get Started" and follow the prompts.

- On your smartwatch, launch the Activity app.
- Proceed by swiping left, and then move to the bottom of the display.
- You can now add a friend by touching on the "**Invite a Friend**" icon and then clicking on a friend.
- You can also delete a friend by touching the friend you're sharing with and then touch the "**Remove Friend**" option.
- Once the friend agrees to your invitation, their activity will appear on your smartwatch, and they'll also see yours. However, if they're yet to accept the invitation, touch their name under the "**Invited**" section of the "**Sharing**" display, and touch the "**Invite Again**" option.

View Your Friends' Progress

- On your smartwatch, launch the Activity app.
- Proceed by swiping to the left, and then move across your friend's list.
- Next, click friends to view their daily statistics.

Compete With Friends

With a little friendly competition, you'll be able to stay motivated. In order to compete against a friend, you must close a certain percentage of your Activity Rings. Each day, you earn one point for each percent of your rings that you add. Each day you can score up to six hundred points, for a weekly total of four thousand two hundred points. The contest lasts for seven days. The final tally of points determines the winner. Some alerts notify you if you're above or behind your rival—and the score—during a competition.

- On your smartwatch, launch the Activity app.
- Proceed by swiping to the left, then touch on a friend, and touch **Compete**.
- Next, touch the **Invite [friend's name]** option, and hold on for your friend to accept.
- Otherwise, if you get an Activity sharing alert informing you that your friend closed their rings or increased their goal, you can swipe down and touch the **Compete** option.

Change Your Friend's Settings

You can manage and change your friend's settings. Launch the Activity app on your smartwatch, swipe to the left, and touch a friend. Use the Digital Crown to move down, and perform any of the actions below:

- To silent alert for the friend: Touch the **Mute Notifications** option.
- To hide your activity: Touch **Hide my Activity**.
- To delete the friend: Touch the **Remove Friend** option.

Accept An Invitation

Once your friend sends you an invitation to share an Activity or contest, an alert will come up on your smartwatch. Touch either "**Accept**" or "**Ignore**." However, you can also check the Fitness app on your iPhone if you don't see the alert on your smartwatch.

- Bring out your iPhone and launch the Fitness app.
- Next, click on the **Sharing** tab.

- Then, click on the Account icon  at the top of the display.
- Next, click either the **Accept** or **Ignore** option.

Change Fitness Goals

- On your smartwatch, launch the Activity app.
- Move down to the bottom and touch the "**Change Goals**" option.

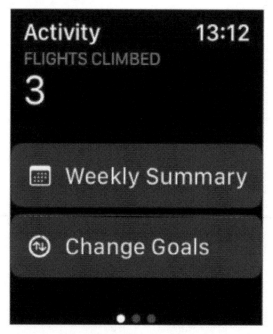

- Then, touch either the "+" or "-" icons to boost or lower your Stand, Move, or Exercise goals.

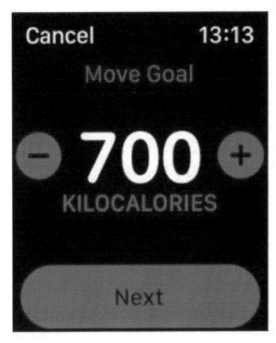 

- Next, touch **"OK"** to save.

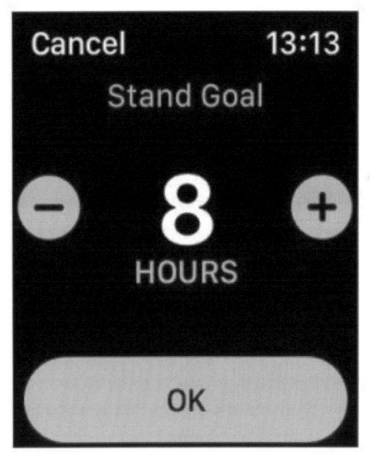

Calm Yourself With The Breathe App

The Breathe app has been renamed "Mindfulness," with new animation to help you center and calm as you breathe. The "Reflect" choice provides a minute-long session for reflection.

You can use your Apple Watch's Mindfulness app  to practice deep breathing or to think about a specific activity or thought.

Start A Reflect Session

- On your smartwatch, launch the Mindfulness app.
- Next, touch the 3 dots icon .
- Proceed and touch **Duration** to change the duration of your session.
- Then, touch **Reflect**.
- Touch **Begin**.
- You can close the session earlier than anticipated by swiping right and tapping **End**.

Start A Breathe Session

- On your smartwatch, launch the Mindfulness app.
- Next, touch the 3 dots icon.
- Proceed and touch **Duration** to change the duration of your session.
- Then, touch **Breathe**.
- You can close the session earlier than anticipated by swiping right and tapping **End**.

Adjust Mindfulness Reminders

- On your smartwatch, launch the Settings app.
- Next, touch on **Mindfulness**.
- Proceed by touching the "**Start of Day**" or "**End of Day**" option. To set a custom reminder, touch the "**Add Reminder**" option.

- You can choose to change the haptics of your reminders or touch the "**Weekly Summary**" option to get an alert on Monday with the former week's Mindfulness activity.

View And Monitor Your Mindful Minutes

- Bring out your iPhone and launch the Health app.
- Then, click on **Browse**.
- Next, click on **Mindfulness**.
- Proceed by tapping on your **Mindful Minutes** to see an illustration of your minutes for the day. Also, you can view your Mindful Minutes for a different timeframe by clicking on the tabs over the top of the chart.
- You can also see your **Mindful Minutes** in the Fitness app.

Set A Session Duration

- On your smartwatch, launch the Mindfulness app.
- Next, touch the **More Options** icon ● ● ●.
- Then, touch **Duration**, and click a duration.
- Proceed by clicking a time between 1 and 5 minutes.

View Heart Rate During Mindfulness Sessions

After your **Breathe** or **Reflect** session is over, your heart rate will be shown in the **Summary** interface.

Also, you can see the overview later on your iPhone:

- Launch the Health app.
- Click on **Browse**.
- Then, click **Heart**, and touch **Heart Rate**.
- **Next, click on** Show More Heart Rate Data.
- Proceed by swiping up and clicking on **Breathe**.

Use The Breathe Watch Face

You can add the Breathe watch face to your smartwatch to quickly access your mindfulness sessions.

- While on any watch face, start by touching and holding onto the screen of the current watch face.
- Proceed by swiping left and to the end.
- Next, touch the New icon "+."
- Proceed by turning the Digital Crown and touch on **Breathe**.
- Next, touch **Add**.

Then, touch the watch's face to launch the Mindfulness application.

Monitor Stocks And Shares

Keep tabs on happenings in the stock market with the Stocks app. Customize the watch face by adding the Stock complication, launch the Apple Watch app on the iPhone > My Watch > Stocks and pick a default stock. The stock selected will be shown on the watch face by default. The indicators you want to see on the watch face like current price points change, percentage change or market capitalization can be added using the Apple Watch app on iPhone > My Watch > Stocks and pick the desired indicators.

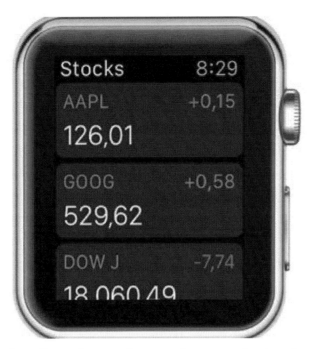

To manage the stocks displayed on the Apple Watch, open the Stocks app > Add stock, dictate the app's name and click 'Done'.

Track The Time Around The Globe
World Time

This dial allows you to track time in 24 time zones at a time. The locations around the external dial represent different time zones, and the internal dial displays the current time at each location. When you touch the globe, focus on the time zone you are in, which is also indicated by the 6 o'clock arrow.

Icons of the sun and moon represent the sunrise and sunset wherever you are, and the light and dark areas of the world reflect the night and day moving across the earth.

With watchOS 8, Apple Watch supports two new types of exercise: Pilates and Tai Chi.

Flexible features: analogy or digital time

Available complications: Activity • Alarms • Astronomy (lunar phase) • Sound books • Blood oxygen • Calculator • Calendar (current date, your time) • Camera remote control • People Search

Set Up A Stopwatch

On your Apple Watch, open the "Stopwatch" app. You can add the Stopwatch complication to your watch face, as outlined in Chapter 6.

- On your **Apple Watch**, touch the side button.
- Swipe to scroll and tap the "Stopwatch" app.
- To start the Stopwatch, tap the green "Start" button in the bottom right corner of the screen. This button turns red and changes to "Stop" to indicate the stopwatch is running.
- Double-tap the Stopwatch screen to switch between analog, digital, graph, or hybrid modes.

Laps

The screen's white switch in the bottom left corner is used to record laps.

- While using the "Stopwatch" app on your **watch**, tap the white switch. In the analog view, tap the green button to start the stopwatch.
- Double-tap the white button in the bottom left corner of the screen. The tab bar shows "Lap" and "Start."
- Tap "Start."
- While the stopwatch is running, tap "Lap" to record a new lap. The screen displays L1, L2, etc.
- Swipe up to return to the main screen. Swipe up again to return to Lap View.
- Tap "Stop" when done.
- Tap "Reset" to begin a new set of recordings.

Monitor The Weather
Check The Weather Forecast

- View the actual temperature and weather conditions for the day: On your Apple Watch, open the Weather app. Click a city, after which click the display to cycle through hourly rain, weather, and temperature forecasts.

- View information about the UV index, air quality, wind speed, and a 10-day forecast: Scroll down by tapping a city.

To go back to the list of cities, tap the left arrow in the top-left corner.

Note: Not all regions have readings for air quality.

Include A City

- On your Series 7 device, open the Weather app.
- Tap Add City at the bottom of the list of cities.
- Enter the city name manually (Apple Watch Series 7 only), or use Scribble or dictation.

Scribble can be accessed on the Apple Watch Series 7 by swiping up from the lowest part of the screen and then tapping Scribble.

Please keep in mind that Scribble is not accessible in all languages.

- After tapping **Done,** tap the city name from the list of outcomes.

The Weather app on your iPhone displays the same cities, in the same order, as those added to the Apple Watch's Weather app.

Eliminate A City

- On your Series 7 device, open the Weather app.
- Swipe the city you wish to delete to the left in the list of cities, then tap X.

Your Series 7 device and iPhone are now devoid of the city.

Select A Default City

- On your Series 7 device, open the Settings app.
- Select a city by tapping **Weather**, then **Default City**.
- Additionally, you can access the Apple Watch app on your iPhone and navigate to Weather > Default City by tapping My Watch.

If you've incorporated weather to the watch face, the current conditions in that city are displayed.

Consult The Weather Forecasts

A notification could show up on the Weather app when an important weather event is predicted.

Navigate Your Watch With Siri

No Apple device would be complete without Siri, the voice-controlled virtual assistant. Siri is integrated into your Apple Watch, and you can use it just like you do on your iPhone. The small display size on an Apple Watch can make it difficult to carry out some tasks, but Siri can easily carry these out for you.

To activate Siri on your Apple Watch, all you need to do is:

- Say "Hey Siri"
- Or, press down and hold the Digital Crown.

- You will see the Siri icon appear and wait for you to make a request.

If this doesn't work, you can check to see if Siri is turned on for your Apple Watch:

- Press the Digital Crown to see your apps on the Home Screen.
- Locate the Siri app and tap the icon.
- Make sure that all the options—Hey Siri, Raise to Speak, and Press Digital Crown—are toggled to the position.

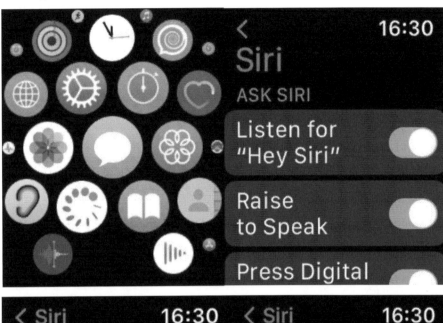

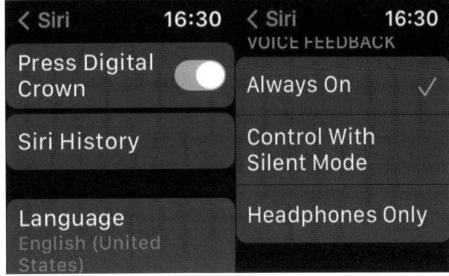

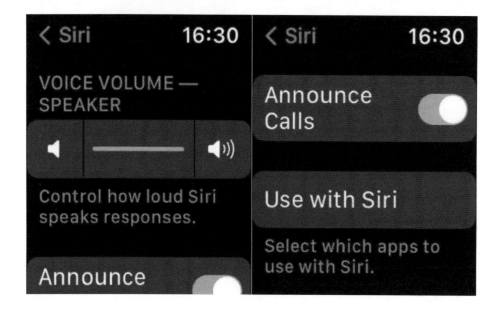

You can use Siri to make basic requests:

- "What is the time?"
- "What is the weather going to be today?"
- "Open the Photos app".
- "Start a 10-minute timer".
- "Tell my wife I will be home at 7 pm".
- "What is my heart rate?"
- "What song is playing?"
- "Convert inches into millimeters".
- "Start tracking my exercise."
- "Do a rap" or "Sing a song".

You can also ask Siri how it can help you further.

Messaging

Siri makes it easy to write, reply, and send messages on your Apple Watch. All you need to do is give the right command:

- Call someone: Siri can start a phone call with any contacts in your phonebook.
- Call a number: If you do not have a contact saved in your phonebook, just read the number to Siri, who will dial it for you.
- Redial my last call: Instantly connect with the last person you spoke to.
- Text someone: Tell Siri to text one of your contacts, and it will open the message dialogue, before asking you what you want to say. Read out your message to Siri and then say "Send."
- You can also include your message directly in your Siri request, e.g., "Tell John I will meet him at the corner." Siri will send John a message, saying "I will meet you at the corner."

- You can also ask Siri to read your last message or the last one from a specific contact.

Calendars And Timekeeping

Siri is seamlessly integrated into your Apple Watch's clock, schedule, and calendar functions. Ask Siri to do anything from setting the alarm to calculating how long until your next doctor's appointment:

- "What is the time?"
- "What is the next appointment on my calendar?"
- "What is the time in Minnesota?"
- "What time is sunrise tomorrow?"
- "When is Sue's birthday?"
- "When is my next appointment with Dr. Schafer?"
- "Set an alarm for 6:30 tomorrow morning."
- "Start a stopwatch."
- "Create an event for August 5th at noon."
- "Make a note."
- "Remind me to buy toothpaste next time I visit the store."

Navigation

You can use Siri to navigate using Apple Maps or Google Maps if you have the apps installed on your iPhone and synced with your Apple Watch. You can ask Siri where a location is, or to give you directions:

- "Where is the nearest grocery store?"
- "Direct me to a gas station."
- More From Siri
- Siri is not only a useful assistant; it can perform many tasks you may not have thought of.

- Translation: Use Siri to translate phrases from one language into another. This makes traveling much easier as you can communicate with people in their own foreign language.
- Apple Pay: Use Siri to send money to one of your contacts.
- Turn settings on or off: Siri can turn on Do Not Disturb, Airplane, or Focus, as well as most other settings on your Apple Watch.
- Flip a coin: Use Siri to flip a coin in a game of chance.

Download Third-Party Apps

- Open the App Store application⊙ on your Watch.
- Rotate the Digital Crown to view specific applications.

Touch any category or touch **See All** below a collection to view more applications.

- Touch **Get** if you want to get a free application. Click on the price to purchase an application.

If a Download icon⊕ is displayed on the application instead of the price, it means that you have already bought the application and can download it again for free.

Click the Search field at the top of your watch display to find a specific application, then type or use dictation or Scribble to type the application's name.

Manage Your Installed Apps

To remove an app from your Apple Watch, simply hold down the Home Screen, click Edit Apps, and tap the X. Unless you delete it from your paired iPhone, it will remain there.

Additionally, you can swipe the app left in the list view and afterwards click the Trash button to delete it from your Apple Watch.

Not all apps on your Apple Watch can be deleted.

Modify The App's Settings

- On your iPhone, launch the Apple Watch app.
- To view the apps you've installed, click **My Watch** and then scroll down.
- To modify an app's settings, tap it.

Certain restrictions enabled in Settings > Screen Time > Content & Privacy Restrictions on your iPhone also apply to your Apple Watch. For instance, if you deactivate the Camera on your iPhone, the Apple Watch Home Screen will be devoid of the Remote Camera icon.

Chapter 5:

Apps Installed In Your Apple Watch

44 Applications (apps) came with your Apple Watch Series 7. The functions of the whole app will be briefly summarized below for you to know their uses and recognize their icons whenever you see them on your watch.

Subsequently, I will mention the name and the icons of the various apps to help you know their individual functions and different features embedded in each of the apps.

Settings App

 This is primarily used to find applications and activate the general features that can be benefited in the apps. You can only use the settings app to access the various settings in all the available apps on your apple watch.

As the name implies, "settings" is the brainbox of all the apps. It can be used to set all the apps' features and solve all problems that can affect the adequate functions of the entire apps.

Apple Store App

 The Apple Store app will enable you to get different compatible apps that can improve your daily activities in terms of communication or conversation, social interaction (i.e. Facebook, Twitter, Instagram, etc.), health, fitness, and time management on your watch.

This app can also enable you to get new free or purchasable apps from Apple Store. More so, you can download/install third-party apps available on your iPhone. However, all the installed apps from Apple Stores can be seen on the Home Screen of your Apple Watch.

Activities App

 The Activity app will enable you always to keep your body healthy and fit by monitoring your daily steps and standing duration. It measures every movement you have taken and the length of the physical exercise you have done so far.

It shows three different colored rings (i.e. red, yellow, and green) that explain the volume of calories used for your long sitting, regular movement, and adequate exercise you have performed daily to complete the three colored rings.

Alarms App

 The Alarm App will enable you to set an important time on the watch. The alarm will produce a continuous sound to remind you of the time set until you stop it.

More so, you can also ask Siri to set a continuous alarm for any specific time (e.g. 1 a.m, 1 p.m., etc.). Immediately, Siri will carry out the request on your watch and tell you that the Alarm has been sent to your requested time.

Snooze is available to stop the alarm sound for a while. After a few minutes, the alarm sound will automatically continue.

Audiobooks App

If you have enough storage space, your Apple Watch can effectively sync **Audiobook** from the Apple Book Store. There is a chance of listening to more audiobooks.

You can fully listen to 5hour of every Audiobook you have added, and it will be automatically downloaded into your watch.

You will need to connect your Apple Watch to Power to help Audiobook sync into it.

Blood Oxygen App

 The Blood Oxygen App measures the volume of Oxygen available in your red blood cell. It shows the reading in percentage to inform you of the amount of oxygen transported from the lung to your body's general blood circulatory system.

This application is not available in all regions, but where it is available, you will see the App icon functioning. But, the reading is not recommended for medical measuring devices.

Calculator App

The calculator app on your watch will enable you to calculate simple mathematical calculations.

If you are a banker, mathematician, or business person, you do not need to look for any ordinary calculator to calculate your business income, stock, summation of goods or commodities, and profits accrued daily.

Calendar App

14 The Calendar App on your Apple Watch will enable you to save both essential events and common events on your watch. On the day or the time of the events, the watch will show you a notification on the screen as a reminder and backup with sound to call your attention. The Calendar will show you the entire event on your watch and iPhone.

Camera App

The Camera app on your Apple Watch can clearly view many distal pictures you have taken with your iPhone.

What do I mean by distal pictures?

You have taken these pictures from a far distance, not a close shot. You can determine the perfect distal picture on your Apple Watch before you take the picture with your iPhone.

More so, your watch can be used to perfectly set the **Shutter Timer** to give you enough time to bring down your wrist and normalize your head if you are to be in the shot.

For your Apple watch to perform camera remote, it should be in the same Bluetooth range as your iPhone, that is, close to 33feet or 10 meters in the distance.

Compass App

This Compass App will show you the direction in which the screen face of your Apple Watch is looking. It will also show your present elevation and location (i.e. where you are).

You will always see your location at the watch face's top left. If you remove the app from your iPhone, the app will be instantly removed from your Apple watch.

Cycle Tracking App

Cycle Tracking App is primarily useful for women not in the menopause stage to track their menstrual cycle. This app will provide an accurate record for women that cannot specifically predict their next menstrual period.

Females can properly record the various symptoms like headache, cramp, discharge, etc. that occur before, during, or after their menses.

With your menstrual logged history, the Cycle Tracking App will alert you to the next menstrual cycle and possible ovulation period.

ECG (Electrocardiogram) App

ECG app will enable you to determine the electrocardiogram of your heart and the heart rate that will be measured in **Bit Per Minute** (BPM) on your Apple watch series 7. You will be given where to record the symptoms of your health discomforts. Ensure that you are using iPhone 6s and above, and upgrade the iOS to the latest version.

You should also know that this ECG app does not function in every region. But, when you are in a functional location, you will see it active when you use it.

Find People App

The Find People app can enable you to share your location with your trusted friends and close family members.

You can use the Find People app to discover and locate the present position of those people you have previously shared your details with.

The friends and family members using iPhone, iPad, iPod touch, Apple Watch SE, Apple Watch Series 3, 4, 5, and 6 will obviously appear on **Find People Map** for you to know where you can find them. This is possible because they have shared their location with you.

A notification can also be put in place to receive an alert when your friend or family member leaves a particular location and when they get to a different particular place.

Find Devices App

This app can be effectively used on your Apple Watch to locate your lost or misplaced Apple device(s). Connecting all your Apple devices to your Apple ID is very important. The use of one Apple ID for all your Apple devices will link them together.

Use Find My App on your any of these Apple devises like the iPhone, iPod Touch, iPad or Mac.

Find Items

 This App can be used to locate the third-party lost or misplaced item or your AirTag connected with your Apple ID.

Heart Rate App

The Heart Rate app carefully monitors the condition of your heart rate. It confirms the normality or abnormality of your heart.

The condition of the heart determines the general wellness of your body. If your heart is healthy you will surely be capable of doing all the stressful activities or performing your daily activities without experiencing any difficulty afterwards.

This will also enable you to check your heart rate during a workout, relaxation, breathing, trekking and others. You can consider taking a new reading at your preferable period.

Home App

The home app will give you a renounced safety technique to regulate and automatically manage your vital HomeKit-enabled accessories like thermostats, locks, window shades, cameras or CCTV, electric lights, smart plugs... and many others.

However, this makes it easy to manage all your HomeKit-enabled accessories on your Apple watch without stress. You can also use Siri to turn off any automatic gadget.

Mail App

 The Mail app will enable you to read all your emails, create/compose, or reply message(s) and send mails to contacts on your Apple watch.

Be informed that all the contacts on your iPhone can be equally found on your Apple watch. All you have to do is to tap on the contact or favorite as you normally do on your iPhone.

More so, you can delete unwanted/read emails on your Apple watch directly.

Maps App

Maps App uses your location and environment to determine where you are. It can be used to know the present place you are.

Similarly, you can ask Siri, "Where is the nearest fueling station"; instantly, it will tell you a nearby fueling station around you.

Memoji App

 Memoji App provides an opportunity on the Apple watch to customize/create your own preferred Memoji. During the customization, you will select skin color, freckle, hairstyle, facial appearances, glass... and many others to look like you or the person you like closely.

You can use the customized Memoji to make calls or conversations (i.e.) on iMessage or Watch Face.

Message App

Message app will enable you to receive a text message and to reply to the text message through the text composition, conversion of dictation to text, scribble, or by switching to your iPhone to make your reply.

When a message enters, you will hear a tap or alert. However, it depends on the types of sound notification you have activated in the message settings.

Music App

The music app will enable you to listen to different audio sounds or music songs wherever you are. You can also use the Music app without having your iPhone around you.

News App

News App will provide you with the latest information on what will enhance your life scope, and provide updated news that will benefit you.

You will additionally know every outstanding and essential event around you and worldwide. More so, with the support of Siri, you can get more précised latest news as you requested.

Noise App

Noise app will seriously guide you from unhealthy and dangerous noise in your surrounding by showing the ambient sound level around you.

Your Apple watch could measure the degree of the noise through the microphone and time of exposing yourself to the environment.

As soon as the Noise app has discovered that the Decibel level is unhealthy for you, your Apple watch will start tapping your hand wrist to excuse yourself from where you are.

Now Playing App

 The Now Playing app provides a panel to control audio playback on your Apple watch easily. You can use the "Now Playing" on your Apple watch to control fast-forward, rewind/backward, pause, or stop music, podcast, or an audiobook on your iPhone.

Phone App

The Phone app will enable you to make a phone call or receive a phone call from a caller. This is an amazing feature that makes the Apple Watch inevitable and important to every user.

This implies that you can directly make or answer a call on your Apple watch without using your iPhone. You can use it to access Voicemail or incoming calls or missed calls on your watch. If you do not feel like receiving the incoming call on your watch, you may switch it to your iPhone from your Apple watch.

Photo App

The Photo app will enable you to access all the photos in the photo library of your iPhone. You can view Live Photos or create a Kaleidoscope on your photos.

You can also add a new photo watch face on your Apple Watch. On your Apple Watch, you can also access photo memories, favorites, and past events from your Photo library.

Podcasts App

Podcasts app will allow you to listen to podcasts anywhere without having your iPhone around you. When your Apple Watch is connected with power, all the latest subscribed episodes of Podcast on your iPhone will be automatically added to your watch.

Radio App

 Radio app will enable you to connect to different stations of different frequencies that provide amazing music, informative news, and the three Apple stations (i.e. Apple Hits, Apple Country, and Apple Music) where you will be listening to the educative and creative interview, latest music and positively life-building information.

Reminder App

 The reminder app will enable you to set important events and accurately remind you of every essential program you have composed and saved on your either Apple watch Reminder or iPhone.

You can make a reminder on your iPhone and access it or edit it on your Apple Watch.

Remote App

 Remote App is effectively active in regulating Apple TV when it is connected to the same Wi-Fi network. It will enable you to access the Apple TV menu to select your preferred option; play and pause selections are available on your Apple Watch to either continue watching or stop the Apple TV.

Shortcuts App

This Shortcuts app will enable you to quickly turn on different features you have made on your iPhone shortcuts with a single tap. Home could be instantly accessed, and you can make more than twenty-five playlists in your Shortcut app.

Sleep App

Sleep app will enable you to plan for your adequate bedtime daily to satisfy nature necessities that will help your healthy living. Put your Apple Watch on your wrist when you are going to bed for it to track your planned sleep.

However, when you awake, tap on the Sleep app to know the duration of your sleep from night to the following morning. You can track your sleep for two weeks consecutively. Ensure that your Apple Watch is fully charged before you go to bed.

Stocks App

Stocks app will enable you to access and monitor the information about all your Stocks on your Apple Watch. More so, the Stock app will show all the available Stocks on your watch with the ability to add or remove any unwanted stock.

You can further ask Siri for help, e.g. by saying, "What is today's opening price for Amazon stock?"

Stopwatch App

Stopwatch app can enable you to time your activities for a maximum of 11 hours and 55minutes. Stopwatch could be used to ensure precession and perfection of programs. Therefore, you can use a lap or split time stopwatch to organize, correct and adjust any irregularity. You can use the ***start button*** to commence the timing or use stop to end the timing

TimerApp

 Timer app is efficiently useful to track time for your activities within 24 hours. It could be used in sports activities, laboratory analysis, construction businesses, presentations, jingles, advertisements... and many others.

Timer app has a Start button to start the time tracking from a second to 24 hours and you can tap on the Cancel button when you are done.

Voice Memos App

Voice Memos app will enable you to record your voice or any external presentation with your Apple Watch. The recorded always shows the recording duration, the time you recorded your voice or others.

It has activation buttons to tap for the recording to start or end. The recorded voice will automatically sync with your iPhone or other Apple devices you are using.

Walkie-Talkie App

Walkie-Talkie App provides a lovely communication method to talk with another acceptable Apple watch user. It exactly functions like the original Walkie-Talkie: You have to press a button to talk and later release it to hear the response. On your Apple Watch, you will touch and hold the button on the screen to talk. The Walkie-Talkie app uses Cellular, Wi-Fi networks, or Bluetooth to enjoy uninterrupted communication.

Wallet App

Wallet app will give you optimal benefits of enjoying a safe, confidential, and simple method of performing payments on your Apple Watch without any difficulty. You can use your Wallet app to make contactless, person-to-person, and transit card payments.

Weather App

 Weather app is a fantastic feature that forecasts the climatic condition of the day. It helps every individual to know how to prepare their day ahead. You can also ask Siri to tell you what the forecast for the next day will tell you details of the day's weather.

Workout App

 Workout app is very helpful to everyone that performs daily workouts to achieve a healthy body fit and good living. It gives a provision to fix time, distance to cover, and calories. Your Apple Watch will track your advancement, notifying you of the distance you have covered so far, and produce the final result in brief.

Subsequently, you can use the Fitness app on your iPhone to completely go through the total Workout details.

World Clock App

 World Clock App will enable you to know the exact time in the various cities or countries around the world.

Chapter 6:

Best Apps To Download

While the Apple Watch is full of features, much of what it can do depends on which apps you choose. The following is a list of some of the best apps, both stock and third party, to help you get the best out of your Apple Watch.

Free Apps To Download
CityMapper
This public transport app gives you directions on navigating selected cities, including New York, Paris, Manchester, Barcelona, and London. It provides directions and times for subways, trains, and buses, cycle routes, and details of bike hire companies. As soon as the journey is set up, you will get step-by-step notifications on your Watch.

Outdooractive
This app is a great fit for you if you love hiking and walking in nature. It is much like most other fitness tracking apps—it can record and display your activity duration, route, ascent, descent, elevation, position, heart rate, and burned calories. However, Outdooractive is unique because it uses GPS to help you navigate on trails using audio cues so that you can safely explore the wilderness without fear of getting lost or taking a wrong turn. The app helps to direct you by vibrating as you approach turns and when you go off-route. You can also see a direction arrow on your watch that points in the direction you need to travel. The arrow will point 50 meters ahead of you so that you can identify turns before you get to them.

You can use Outdooractive without your iPhone, which makes it a great choice for all-weather activities where your iPhone may be damaged by water, dust, and dirt. Unfortunately, the navigation features will be limited if your watch is not connected to your iPhone. You are able to download maps to your Apple Watch beforehand to get access to some of the best features while not connected to your iPhone. You will be able to see a map with your location on your watch, and you can easily pan or zoom in and out to see different locations. Switch between different map modes to see different aspects of the environment, like topography, satellite imagery, or standard view.

Calm

Calm has won several awards, including the Best of 2018, 2017 App of the Year, and the Happiest App in the World. It is ranked as the top app for sleep, meditation, and relaxation, and you should try it if you struggle to manage your stress levels or sleep patterns.

The app provides guided meditations, breathing programs, stretching exercises, relaxing and calming music, and stories to help you sleep. You can access hundreds of different programs geared towards your specific needs, experience level, and available periods. The guided meditations have different focuses, ranging from deep sleep to happiness and gratitude, self-esteem, relationships, calming anxiety, and mindful walking. The bedtime stories are narrated by some of the world's most beloved voices, including Stephen Fry and Matthew McConaughey. One of the most understated features is the soundscapes, which help to immerse you in ocean waves or near a campfire as you meditate.

Calm is a subscription-based service that offers a free trial period, and it can collect and save all your meditation data and sleep sessions directly into the Apple Health app. It integrates seamlessly with your Apple Watch to provide different sessions and monitor your health stats.

Headspace

Headspace is quite similar to the Calm app but stands out in how it works with the Apple Watch, offering you meditations and breathing exercises right on your wrist, whether or not you have it installed on your iPhone. The goal of the Headspace app is not only to help you with your mental health but also to equip you with tools and strategies so that you will always be able to deal with complicated feelings without relying on the app. Each day a new topic is selected as the focus, so you will never get bored with the meditations.

Headspace offers sleep, guided meditations, and a Move Mode, which you can use to help relieve tension in your body by following a series of very basic movements and intentions. One of the best features is the SOS meditation—short three-minute sessions to help you pull yourself together in no time—which is great for difficult situations with family or at work when you feel overwhelmed and need a helping hand. You can also take advantage of the focus-enhancing music that allows you to concentrate and be more productive. This is also a subscription-based service with a few payment options and a free trial period.

Sleep Cycle

This is a highly-rated sleep app that integrates with your Apple Watch. Analyze the quality of your sleep by using the accelerometers and gyroscopes in your Apple Watch to see how restless you are during the night and generate detailed reports about your breathing rate as you sleep. It features smart alarms, where you can set wake windows. For example, if you set the alarm for 6:00 am with a 30-minute sleep window starting at 5:30 am, the app will wait until it detects you are entering a light sleep phase before it will begin to wake you. This helps prevent you from waking up during deep or REM sleep, which can be stressful or frustrating and helps you have a more peaceful and productive day.

Sleep Cycle also offers an extensive library of music, meditations, and stories that have been carefully curated to help promote sleep by reducing anxiety and stress and allowing you to relax. Each morning you can access your sleep analysis record, and the app can help correlate external and lifestyle factors to help you figure out what may be impacting your sleep quality.

Spotify
Free to download; monthly subscription required ($9.99)

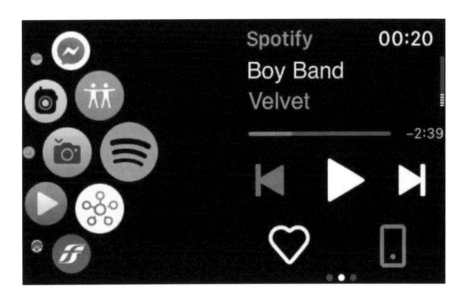

You have likely heard of Spotify if you are not already using it. It is one of the world's most extensive music streaming services, providing you with music on-demand directly to your devices. Spotify can be downloaded for almost any device, including your iPhone, Apple Watch, iPad, laptop, desktop, and even your TV or PlayStation.

Though Spotify is a subscription-based service, you can also use it for free, though some of the features will be limited, and you will occasionally have to listen to ads.

Spotify gives you access to millions of songs from any artist or album you can think of. You can also listen to podcasts, take advantage of the carefully curated charts to see what is trending and popular, and find new genres to pique your interest. Using Spotify, you can also find playlists compiled by the Spotify team and members of the listening community to suit any mood or activity. Find playlists for a lazy Sunday afternoon, a long drive filled with nostalgia, or something more upbeat while you finish some work around the house.

Spotify works directly through your Apple Watch, where you can see important information about the track you are listening to, such as the title, artist, and remaining time left. The control playback buttons are simple and easy to use, allowing you to pause, play, or skip forward or backward. You can also add or remove tracks to your favorites simply by tapping the heart icon and using Spotify connect to play the audio out of your nearby Bluetooth devices, which you can easily switch between.

Swiping right on your Spotify main screen will take you to the Recently Played menu, where you can find a list of all the tracks and podcasts you have listened to most recently. You can also access your Spotify Library and Downloads here. Using your Apple Watch, you can download about 10,000 minutes of listening material so that even if you don't have your iPhone nearby, you can still access your music. This is an excellent feature if you want to go for a walk with your watch but not your iPhone.

Waterful

This cute and quirky app features a purple octopus who acts as your hydration coach to help you drink enough water every day. The app sends reminders to your Apple Watch to make sure you drink and track how much water you get through each day.

Your daily water goal is set based on your gender, weight, height, and age, though you can also customize your goal. It can also consider other factors like any medications you are prescribed, your activity level, and even the weather to ensure you stay hydrated. Use the app on your watch or on your iPhone to log everything you drink and participate in different challenges that help you build better and healthier habits.

GoPoop

This app may not be right for everybody, but your digestive system also deserves attention. Your digestion can tell you a lot about your health, and it is often one of the first signs that something may be wrong. With the focus on fitness, nutrition, hydration, sleep, and heart health, why not spend some time understanding your poop habits better?

With GoPoop, you can create a logbook every time you visit the bathroom, and record different features like regularity, shape, color, constipation, etc. You can use your Apple Watch to swipe through the other characteristics; all you need to do is tap the ones that best describe your poop. With this information, the app will analyze your health and advise you to improve your digestion through diet and lifestyle changes. The app uses the Bristol Stool Scale, a world-renowned tool used to analyze poops and figure out how to treat different bowel and gut conditions.

The app is entirely free, and you can install a companion app on your Apple Watch so you can log your poops at any time.

WebMD: Symptom Checker

If you have ever looked up your medical symptoms online, you were probably directed to one of WebMD's thousands of well-researched and informative articles. They are one of the largest online human health and wellness content publishers. There is also a WebMD app that you can download onto your iPhone, with some convenient features you can set up on your Apple Watch.

The WebMD Symptom Checker app is the place to check any of your symptoms, create reminders for your medications, view allergy alerts, and receive the latest news and updates on health-related topics.

The symptom checker feature allows you to input any number of symptoms you may be experiencing, and the app will suggest any possible conditions or issues that may be related. From there, you can read the thorough information to determine whether or not you may need to visit a doctor or if there are any treatment plans you could try at home first.

With the medication reminders, you can input all of the information about your medications, dosages, precautions, and side effects and receive reminders whenever it is time to take them. You can see your daily medication schedule with any important instructions to keep in mind, and you can also see a picture of the pills so that you never get them mixed up. You can sync your reminders straight to your Apple Watch to view them anytime.

To set up medication reminders, you will need to use the app on your iPhone, where you can select your medications from a list and choose an image from a preset list to accompany them. You can also add custom medications and add your own pictures if you cannot find the correct information from the provided list. Next, select what time you need to be reminded to take the medication. These notifications will now be sent to your wrist, where you can snooze or dismiss the reminders. You can view the reminders by glancing at your Apple Watch, where you will see your upcoming doses and the scheduled time for taking them. When the next dose is due in two hours, you will see a count-down timer instead of the scheduled time. You can also mark doses as taken, missed, or skipped; the app will keep a detailed record that you can refer back to at any time.

The allergy tracker is helpful for keeping track of pollen counts in the air, dust, pollution, mold spores, and other allergy forecasts. You can set your location to view the levels in your area and get alerts when levels in your area are particularly high.

Mango Health

Mango Health is an app that reminds you to take your medication and can give you essential information about different drugs, including their effects and interactions. The app has an easy-to-understand design that helps to simplify your health.

You can set up reminders on your Apple Watch to notify you when it is time to take your medications and supplements. Several other reminders combine the features seen in some other apps that have been mentioned, such as reminders to drink water, a mood tracker, a food and nutrition tracker, a step counter, and alerts for checking your blood glucose levels or blood pressure.

You can easily find important information about all the drugs you are prescribed in the app, alerting you to any side effects and interactions with other medications, foods, or supplements. The warnings are all color-coded to know which ones are more important. If you stick to your schedule without missing any doses, the app will reward you with points. As you collect more points, you stand a chance to win exclusive rewards such as gift cards or charity donations, though this is only available in selected regions. Mango Health can also help you manage your doctor's appointments and ensure you get your next prescription before running out of medication.

Overall, Mango Health is a great app to make your medication regime a little more fun, with a design that is easy to look at and intuitive reminders to keep you on top of things.

Thirsty

Thirst is a smart water tracker that functions similarly to Waterful, though the design is more sleek and minimalistic. It is the perfect companion to your other health and fitness apps that can adapt to your activity level and lifestyle to determine your optimal hydration goals. The app can consider weather conditions such as an upcoming heat wave to ensure you drink enough water and can fend off any chance of heatstroke.

Thirst will send you notifications to help remind you to drink water, and you can easily record your daily water intake by logging your drinks. The app aims to get you to drink a lot of small amounts several times throughout the day, rather than simply to try to hit a goal at the end of the day. It can integrate with your Health app to provide a more comprehensive analysis of your hydration needs and help you achieve them always to feel your best.

ElderCheck Now

This minimalistic app is designed to let those who care about you know how you are doing at the touch of a button. It is more of a check-in app than a communication platform. Through

Eldercheck Now, your caregivers can request a check-in from you. You can quickly respond to the request by selecting 'I'm OK' or 'Call me!' directly on your Apple Watch or iPhone. You can add pictures of your caregiver into the app so their face will come up whenever you receive an alert.

ElderCheck lets you schedule check-ins to ensure everybody is safe and healthy before bed, while out on a walk, or after a doctor's appointment. You can schedule check-ins at any time of day and repeat them every day, week, or month.

When responding to a check-in request, your caregiver can see the time that you responded and your location. This is a good safety feature that can alert your caregivers if something has gone wrong and you cannot answer. If you do not respond to a check-in request, the app will continue to send you alerts.

One of the best features of this app is that it can share your heart rate data with your caregivers every time you respond to their check-in request. This helps them monitor your health even if they cannot be by your side.

To get all of the benefits out of ElderCheck, you must have the app installed on your phone and your caregiver's phone, and also make sure to enable all the permissions such as health and location data.

Emergency: Alerts

The American Red Cross Society develops this app to help you receive accurate information and alerts about climate-related hazards. Just set your location and the locations of your loved ones, and the app will alert you whenever there is a risk of severe weather, such as tornadoes, hurricanes, floods, fires, etc. The app lets you enable alerts and push notifications that you can tap to see more information. You can set the alerts to override Do Not Disturb mode so that you can be notified of an emergency even if you are asleep or busy with other tasks. There are 40 different customizable alerts that you can enable or disable and customize the notifications to suit your particular needs.

The app will direct you to the nearest Red Cross shelter if you need to find a safe place during the weather event, and you can also see an interactive map with detailed overlays such as radar, satellite, clouds, rain, wind speed, and snowfall. You can also receive step-by-step guides that help you prepare and stay safe during a disaster if you cannot get to a shelter.

Strava

Strava is one of the most popular fitness apps, becoming famous for its excellent use of GPS tracking data. With Strava, you can map your run, walk, or cycle to see your routes and get all your workout data like distance, pace, speed, elevation, and calories burned. With Strava, you can see these stats at different points along your route—see zones where you walked faster or where a hill slowed you down, and work to continuously improve and push yourself on the same path to bring down your time.

Through the Strava community, you can find new and exciting routes in your area that other people use. Get access to all kinds of maps and trails that you can try and contribute to with your own.

You can also take part in challenges and competitions through the Strava app. There are monthly challenges where you can hit specific goals and get rewarded. Enter competitions based on your age group, fitness level, and region so you can have fun while pushing your limits alongside those at a similar level.

Running and cycling are the primary sports that made Strava the app it is today. However, it has evolved to accommodate many sports, including outdoor running and cycling, indoor running and cycling, standard gym workouts, walking, hiking, swimming, and many more.

Strava uses the heart rate data on your Apple Watch to monitor your heart rate and assess your performance. You can also receive audio cues through your Apple Watch to your Airpods or Bluetooth headset, giving you updates about your walk, hike, run, or cycle, such as the distance covered current and average pace, and reminders to start, stop or pause your workout.

Become a part of the Strava community by sharing your workouts. You can share them with selected friends or a specific club or community. Your friends can comment on your activities, give you tips, and congratulate you on your achievements.

To set up Strava on your Apple Watch:

- Begin by downloading the Strava app onto your iPhone through the app store.
- Open the Watch app on your iPhone and select My Watch.
- Locate the Strava app and tap to install it on your Apple Watch.

You can record activity using Strava on your Apple Watch without carrying your iPhone with you. Make sure to give the app access to your location and health permissions, accept the legal disclaimer, and turn on notifications when you first open the app on your Apple Watch. Then choose which sport type you want to do and select your preferred unit of measure: miles or kilometers. You can also choose to enable auto-pause so that if you take a break during your activity, the app will automatically stop recording.

During a workout, you can view your elapsed time, average pace, split times, total distance, and heart rate on your Apple Watch. You will be able to turn on real-time audio feedback so that your watch will read this information out to you while you are working out. You can find these settings in the Strava app on your Apple Watch by selecting Settings and Audio Cues. You will be able to see a summary of your workout. To finish a workout, all you need to do is press the stop button and finish. Make sure you then choose to Save your activity so that it can be uploaded to your iPhone.

Strava offers a complication that you can add directly to your watch face. This complication is just a shortcut that will take you directly to the app.

Hole19

This app is a must-have for any avid golfer. Hole19 offers you a simple and easy way to keep score during your games. There are free and paid versions of the app, but this one is definitely worth the price.

With the free version, you can access more than 43,000 different golf courses with a map and GPS data of all the holes, hazards, greens, and other points of interest. Download the course you will be visiting to get all the information you will need to beat your score. You can even get accurate distance measurements from the greens' front, back, and center. The app lets you create a digital scorecard to keep track of every stroke you take. You can see GPS distances on every golf course through your Apple Watch and create live leaderboards so that you can compete with your friends and compare your stats as you play.

The paid version comes with all of these features and more. You can access the helpful handicap calculator or challenge your friends to a match. The shot tracker feature is one of the most appealing—it gets precise measurements of your shots by measuring your starting and finishing position and records the data based on which club you used. The paid version can auto-detect when you move on to the next hole and provide a detailed report of your overall performance statistics and highlights. You can also access the full scoring capabilities through your Apple Watch and compare your game to the club statistics.

Best Apps To Purchase
AutoSleep
Cost: $3.99

A sleek and intelligently designed app that builds upon the foundations of the Apple Sleep app to provide you with more significant insights into your sleep patterns. AutoSleep is an automatic sleep-tracking app that will begin monitoring your sleep using the Apple Watch without the need to press any buttons. You can adjust your AutoSleep settings to suit your needs so that the data will be more accurate—tell the app if you are a light sleeper or a restless sleeper, help it figure out when you go to bed and when you wake up, and what kind of lifestyle you lead. You can see detailed reports after each night's sleep to help you develop better nighttime routines and alter your sleeping habits to improve the quality of your sleep.

Auto sleep uses the heart rate monitor on your Apple Watch to determine the quality of your sleep by looking at how much time you spent asleep while lying in bed, analyzing how restless you are, and how your heart rate changes during the night to develop a comprehensive sleep analysis report each morning. If you have a later model of the Apple Watch, you can even use AutoSleep to monitor your blood oxygen levels as you sleep, which can be used by people with sleep apnea and other conditions that affect breathing during the night.

You can add AutoSleep complications to your Apple Watch watch faces to see some of the most interesting and important pieces of information. One of the best features of AutoSleep is the Readiness rating, which is a simple estimate of your "daily readiness" that considers your physical and mental stress levels and the quality of your sleep. Sleep readiness is determined using many different data points from your Health app, including your heart rate, exercise data, and sleep data.

AutoSleep has smart alarms that you can set that help you wake up without being startled by your usual alarm's harsh beeping or annoying sounds. These smart alarms work over a few minutes and take advantage of the haptic vibrations in the Apple Watch to 'tickle' or 'nudge' you every few minutes, just before you are set to wake up. The smart alarm will determine whether you are in a deep or light sleep phase and decide which method will best wake you up.

AutoSleep includes a charged reminder so that you can charge your watch a few minutes before your bedtime, ensuring the battery will last throughout the night. You can also set Day Time or Night Time watch faces that display all the relevant data fields, and you can easily switch between these just like any other Apple Watch faces.

Dark Sky
Cost: $3.99

Dark Sky is a very specific weather app. The iPhone app logs your physical location moment by moment, and the Watch app subsequently tells you when it is going to snow or rain in your location with incredible accuracy. The Watch app also gives a detailed forecast for the next 60 minutes, the following day and for the week ahead, each on a different screen. The latest version has Complications integration, so you do not need to touch the screen to see the weather.

ETA

Cost: $4.99

Similar to CityMapper, this is an app to help you get places. You simply add a location to the iPhone app, save it and then get timings for walking, driving and taking public transport to get to your destination. Some of the watch faces will have it as a complication, showing how long it will take to get there and taking traffic into consideration as well.

Streaks

Cost: $4.99

Streaks is an app to let you determine the most important tasks for you to achieve. You can pick up to six from the hundreds on your iPhone list, and they will show up in the Streaks complications in the Watch app. Each icon comes with its own set of reminders, and it works in harmony with the Health app to help you achieve your health and fitness goals. The name references the fact that you should aim to hit your targets for a "streak," a number of consecutive days.

Audible

Cost: $7.95 per month

More and more people struggle to find the time or patience to sit down and read books; however, audiobooks are the perfect replacement that allows you to reap all the benefits of reading while still doing other tasks. With Audible, you can listen to some of the best-selling audiobooks and podcasts to keep you entertained for hours. Use Audible when traveling, cleaning, walking, or just relaxing.

You can use Audible to access some of the most famous literary works narrated with great care and attention to detail so that the true essence is captured and conveyed to the listener. You can also find all kinds of genres, ranging from crime, mystery, thriller, horror, historical, romance, and western, as well as many non-fiction categories like history, biographies, memoirs, travel, philosophy, religion, and spirituality, self-help, finance, science, and more.

Audible offers you subscription options that give you a free monthly credit that you can spend on any title in their library while also providing massive discounts of up to 80% off premium titles.

Use the Audible app on your Apple Watch by downloading it onto your iPhone and then syncing your library. You can choose which titles to sync to your Apple Watch, helping you save on storage space. Listen to your audiobooks using a Bluetooth device like headphones or a speaker. The Apple Watch app lets you control the playback and allows you to skip forward or backward 30 seconds, so you never miss anything important. One of the best features is the sleep timer, which you can access by tapping the small clock icon. With this feature, you can set the desired time, after which the playback will stop. Many people find audiobooks so relaxing that they use them to fall asleep, ensuring that the playback doesn't continue through the night, causing you to miss out on all the action. You can also speed up or slow down the pace of the narration.

Chapter 7:

Apple Watch Glossary

- **Apple ID:** This is an email address registered with Apple.
- **Apple Store**: This is a retail and online powerhouse that sells Apple products.
- **Apple TV:** Apple's original Mac OS X Tiger-based living room set-top box designed to buy media from the iTunes Store or stream it from Mac or Windows iTunes.
- **Google:** This provides net, also for Youtube and other apps.
- **iCloud:** It's Apple's online service that is used to replace MobileMe.
- **iMessage:** This SMS or MMS is used to send free text and multimedia messages from one iOS device to another over cellular and Wi-Fi networks.
- **iOS**: iOS is an Apple mobile operating system that powers Apple products.
- **Wi-Fi:** Wireless networking.'

Gesture Terms:

- **Tap Action:** You will gently place your finger on the screen to touch the Apple watch-sensitive screen.
- **Double TapSwipe:** You may be asked to double-tap your watch screen to stabilize or highlight the app or enlarge the view by using a finger to make a quick and fast two taps on the task on your watch screen.
- **Drag Action:** You will need to place your finger on the screen and keep moving it without lifting it.
- **Swipe Left or Right:** When you are asked to swipe left, that means you should move a finger from the right side of the watch face screen to the left, while swiping right is the opposite action of swiping left.
- **Press or Turn the Digital Crown:** I have previously explained the general function of the Digital Crown, that is, it functions as the Home Button when you press it, and it can be used to search for an option from the different available options in your Apple watch by either turning the digital Crown in the up or down direction (i.e. clockwise or anticlockwise).
- **Press the Side Button:** It can enable you to access favorite contacts, mail, power switch or preserve, and different apps on your Apple watch.
- **Touch-Hold the Screen:** This will enable you to edit your watch face or launch the control center. The Touch-Hold means you are to use a finger to press the screen for a second or two seconds. When you press and Touch-Hold the screen, the clock pattern on

the watch face will reduce in size. When you finish editing or selecting, tap the screen once to restore the edited or selected clock pattern to normal size.

- **Zoom In:** This can be done when you move your two convenient fingers far from each other on the watch face screen to increase the image size on the screen.
- **Zoom Out:** This is the opposite action of Zoom In. It can be done when you move your two convenient fingers together (i.e. Pinch) on the watch screen to reduce/decrease the image size on the watch screen.

Conclusion

Apple Watch is, undoubtedly, one of the most exciting devices to come out of the Apple stable in recent years. It isn't just a watch; it is an extension of your iPhone, a device that makes things easier and more interesting. Many of us wondered how Apple would pull it off, how they would get so much into such a small device. They promised us a Watch to be proud of, and they delivered on that promise.

Let's face it. Smartwatches are nothing new; they have been around for many years. There was no doubt about the excitement that the announcement about the Watch caused, and it appears that, once again, the Cupertino Company has come through with the goods. In the months ahead, there will be many newer features and many newer capabilities for the Apple Watch. Enjoy what you have now and join in with the new revolution.

Consult this manual whenever you need to remind yourself of a particular feature of the Apple Watch and how to use it. Thank you, and I wish you all the best!

Made in the USA
Las Vegas, NV
01 February 2024

85191208R00068